Turbulence and Dispersion in the Planetary Boundary Layer

Physics of Earth and Space Environments

The series *Physics of Earth and Space Environments* is devoted to monograph texts dealing with all aspects of atmospheric, hydrospheric and space science research and advanced teaching. The presentations will be both qualitative as well as quantitative, with strong emphasis on the underlying (geo)physical sciences. Of particular interest are

- contributions which relate fundamental research in the aforementioned fields to present and developing environmental issues viewed broadly
- concise accounts of newly emerging important topics that are embedded in a broader framework in order to provide quick but readable access of new material to a larger audience

The books forming this collection will be of importance for graduate students and active researchers alike.

More information about this series at http://www.springer.com/series/5117

Francesco Tampieri

Turbulence and Dispersion in the Planetary Boundary Layer

Springer

Francesco Tampieri
CNR ISAC
Bologna, Italy

ISSN 1610-1677 ISSN 1865-0678 (electronic)
Physics of Earth and Space Environments
ISBN 978-3-319-43602-9 ISBN 978-3-319-43604-3 (eBook)
DOI 10.1007/978-3-319-43604-3

Library of Congress Control Number: 2016950206

Printed on acid-free paper

This Springer imprint is published by Springer Nature
The registered company is Springer International Publishing AG Switzerland

Preface

The purpose of this book is to give an updated and problematic description of the atmospheric turbulence and the dispersion in the planetary boundary layer, for students, and perhaps to researchers, interested in the atmospheric sciences. Well-settled arguments coexists with topics investigated in the last years or even under discussion: thus, the reader will find differences in the treatment of the arguments that reflect these aspects. The hope is to give a comprehensive view of the physics of the planetary boundary layer, with the certainties and the uncertainties, to raise interest and to stimulate future research.

A couple of observations (which should become suggestions to the reader). The first one: many boring computations are put into the exercises. Most of them refer to analytic solutions of simplified problems. Although numerical modelling is of increasing use, the exact solutions may be a benchmark for models and are useful to better evidence the basic mechanisms and approximations. The student should not overlook this aspect.

The second one: in the Appendix to the Introduction, some data sets are cited, and reference is given to download the data. Note that the list is not exhaustive; other data sets are freely available. The data typically refer to variables averaged over some time interval. Most of the figures of the book are made using such data, so that every figure can be understood as an exercise for the student, who can remake, modify and possibly improve it. (As a general rule for the figures, the data are plotted directly, with various symbols, or binned in intervals of the independent variable and reported in terms of the median and the 10th and the 90th percentiles, with error bars.)

I would like to acknowledge here all who contributed to the genesis of the book: first, all my students who pushed me to build (as far as possible for me) a clear and unifying picture of the topic and the colleagues for discussions and criticism, in particular those who gave me data and special images. Special thanks to Domenico Anfossi, Alessandra Lanotte, Silvia Trini Castelli and Sergej Zilitinkevich, who suggested specific arguments and supplied the proper material. Last but not least,

I must remember two people who addressed my research and thus are in part responsible for the genesis of this book: Ottavio Vittori, who taught me not to cross the road on zebra crossing, and Julian Hunt, who introduced me to the mysteries of turbulence.

Contents

Acronyms and Symbols

CBL	Convective boundary layer
DNS	Direct numerical simulation
FPE	Fokker-Planck equation
IBL	Internal boundary layer
K41	Inertial subrange paradigm, from Kolmogorov (1941)
LE	Langevin equation
LES	Large eddy simulation
LSM	Lagrangian stochastic model
MOST	Monin-Obukhov similarity theory
NSE	Navier-Stokes equations
NWP	Numerical weather prediction
PBL	Planetary boundary layer
RL	Residual layer
SBL	Stable boundary layer
TKE	Mean turbulent kinetic energy
TPE	Mean turbulent potential energy
UBL	Urban boundary layer
UCL	Urban canopy layer
WMC	Well-mixed condition

a	A generic scalar variable; the inertial particle radius, Eq. (5.103)
$\mathscr{B}$	Buoyancy term in the TKE equation, Eq. (3.7)
$\tilde{c} = C + c$	Concentration of the scalar c, Sect. 2.2.4
c_{g}	Geostrophic drag coefficient, Eq. (3.77)
c_{d}	Drag coefficient for velocity in the UCL, Sect. 4.6.1.2
c_{p}	Specific heat at constant pressure, Sect. 2.2.5. For the dry air $1012\,\mathrm{N\,kg^{-1}\,K^{-1}}$

c_U	Drag coefficient for velocity, Eq. (3.32)
C_K	Inertial subrange Eulerian constant, reference value 2, Eq. (2.96)
C_S	Mean (average) concentration from the source S, Eq. (5.2)
C_0	Inertial subrange Lagrangian constant, reference value 6.2, Eq. (2.111)
C_θ	Inertial subrange Eulerian constant for scalars, Eq. (2.116)
D_{ij}	Eddy diffusion coefficient in transport processes, Eq. (5.10)
D_{uu}	Structure function for velocity of order 2, Eq. (2.95)
D_{uuu}	Structure function for velocity of order 3, Eq. (2.97)
E_{ab}	Cospectrum or Fourier transform of R_{ab}, Eq. (2.55)
E_F	Kinetic energy of the filtered field, Eq. (6.39)
E_{ij}	Velocity cospectrum, or Fourier transform of R_{ij}, the covariance between velocity components u_i and u_j, Sect. 2.3.3.8
$E_k = \frac{1}{2}\langle u_i u_i \rangle$	Mean turbulent kinetic energy, Sect. 2.4.3.1
E_P	Mean turbulent potential energy, Eq. (3.12)
E_S	Sub-grid scale kinetic energy, Eq. (6.40)
f	Coriolis parameter, reference value $10^{-4}\ s^{-1}$, Sect. 2.2.1; a generic function
f_E	Eulerian probability density function of the velocity, Eq. (5.3)
f_α	Probability density function of the random variable α, Sect. 2.3.1
$\boldsymbol{F}^{(c)}$	Flux of the scalar c, Sect. 2.2.4
$F_r = U/NL$	Froude number, Sect. 2.2.7
g	Gravity acceleration, reference value $9.81\ m\ s^{-2}$, Sect. 2.2.1, Richardson law constant, reference value 0.6, Eq. (5.34)
h	PBL depth, Sects. 1.3.3, 3.2, 3.5.2, 3.7.4
h_b	Blending height, Sect. 4.4.1
h_c	Roughness element height, Sect. 3.4.1; for an urban canopy, building height, Sect. 3.4.1
h_i	IBL depth, Sect. 4.2.1
h_s	Surface layer height scale, Sect. 3.2
H_t	Scale height of the topography, Sect. 4.3
H	Vertical height scale of the motion

k	Wave number
k_b	Wave number scale of the buoyancy-dominated range, Eq. (3.136)
K_Q	Eddy diffusion coefficient for heat, Eq. (2.85)
K_Δ	Eddy diffusion coefficient in LES, Eq. (6.55)
K_τ	Eddy diffusion coefficient for momentum, Eq. (2.80)
Ku_a	Kurtosis of the variable a, Eq. (2.34)
l	Length in the inertial subrange, Sect. 2.5
l	Instantaneous value of the parcel displacement, Sect. 2.4.6
l_K	Viscous length scale, or Kolmogorov scale, Eq. (2.92)
l_τ	Mixing length, Sect. 2.4.6
l_Δ	Length scale for mixing in LES, Eq. (6.55)
L	Obukhov length in the surface layer, Eq. (3.19)
L_{ab}	Integral length scale independent on direction (for isotropic conditions), Sect. 2.3.3.8
$L_{ab}^{(k)}$	Integral length scale in the k direction, Eq. (2.43)
L_{Ai}	Side of the averaging box, Sect. 2.3.2.3
l_t	Inner layer depth over topography, Eqs. (4.15), (4.22), (4.23)
L_t	Streamwise scale length of the topography, Sect. 4.3
L_c	Penetration length scale in an urban canopy, Eq. (4.55)
L_E	Integral Eulerian length scale in isotropic conditions, Sect. 2.5.4
L_{ij}	Integral length scale for velocity components, Sect. 2.3.3.8
L_r	Scale length of the roughness change, Sect. 4.4.1
L	Global length scale of the motion
n_S	Number of tracer parcels, Eq. (5.1).
N	Brunt-Väisälä frequency, Eq. (2.29). Reference value 0.01 s^{-1}
$p_f = p_a + \tilde{p}$	Pressure of the air, reference value 1013 Kg m^{-1} s^{-2}, Sect. 2.2.1
$P_r = \nu/\kappa_H$	Prandtl number, Sect. 2.2.7
$P_t = K_\tau/K_Q$	Turbulent Prandtl number, Sect. 3.4.2.3
$\mathscr{P}$	Shear production term in the TKE equation, Eq. (3.6)
$\tilde{q}_c = \tilde{c}/\rho_f$	Mixing ratio of the scalar c, Sect. 2.2.4
Q_h	Horizontal kinematic heat flux, Eq. (2.86) and Sect. 3.4.3.4

$\boldsymbol{r}$	Separation vector
R	Gas constant for dry air, 287 $\mathrm{m^2\,s^{-2}\,K^{-1}}$, Sect. 2.2.2
R_{ab}	Covariance between the quantities a and b, Eq. (2.42)
R_{B}	Bulk Richardson number, Eq. (3.21)
$R_{\mathrm{e}} = \mathrm{UL}/\nu$	Reynolds number, Sects.1.2 and 2.2.7
R_{f}	Flux Richardson number, Eq. (3.9)
R_{fc}	Critical flux Richardson number, Sect. 3.1
R_{g}	Gradient Richardson number, Eq. (3.20)
R_{gc}	Critical gradient Richardson number, Sects. 3.7.1.3 and 3.7.3
R_{ij}	Covariance between velocity components u_i and u_j, Sect. 2.3.3.8
$R_{\mathrm{o}} = \mathrm{U}/(f\mathrm{L})$	Rossby number, Sect. 2.2.7
R_{v}	Gas constant for water vapour 641.5 $\mathrm{m^2\,s^{-2}\,K^{-1}}$, Sect. 2.2.5.3
S	Tracer source for WMC, Eq. (5.3)
$S_{\mathrm{c}} = \nu/\kappa_{\mathrm{c}}$	Schmidt number, Sect. 2.4.1
S_{t}	Stokes number, Sect. 5.7
S_{ij}	Mean rate of strain, Eq. (6.4)
Sk_{a}	Skewness of the variable a, Eq. (2.33)
t	Time
T	Integral Lagrangian time scale in isotropic conditions, Sect. 2.5.4
T_{ab}	Integral time scale, Eq. (2.44)
T_{A}	Averaging time, Sect. 2.3.2.2
T_{E}	Integral Eulerian time scale in isotropic conditions, Eq. (2.121)
$T_{\mathrm{f}} = T_{\mathrm{a}} + \tilde{T}$ $= T_{00} + T_0 + \tilde{T}$	Absolute temperature of the air, in Kelvin degrees, Sect. 2.2.2
T_{ij}	Integral time scale for velocity components, Sect. 2.3.3.8
T_{K}	Kolmogorov time scale, Eq. (2.93)
T_{p}	Inertial particle relaxation time (Stokes time), Eq. (5.103)
T_{s}	Characteristic time of the shear, Sects. 5.2.2.1 and 5.6.3.3
T_{v}	Virtual temperature, Sect 2.2.5.3
$\mathscr{T}$	Third-order moments in the TKE equation, Eq. (3.8)
$\tilde{\boldsymbol{u}} = \boldsymbol{U} + \boldsymbol{u}$	Eulerian velocity, Sect. 2.2.1
$u_* = \tau^{1/2}$	Friction velocity, Eq. (3.15)
u_{**}	Velocity scale in the transition sublayer, CBL case, Eq. (3.37)
$\boldsymbol{U} = (U, V, W)$	Mean velocity, Sect. 2.4.1

$\boldsymbol{U}_g = (U_g, V_g, 0)$	Geostrophic wind, Sect. 2.4.1.1
U_U	Upstream (unperturbed) velocity, Eq. (4.3), Sect. 4.3.2
U	Global velocity scale of the motion
$\mathrm{U_l}$	Velocity scale in the inertial subrange, at scale l, Sect. 2.5
$\boldsymbol{v}$	Lagrangian velocity of a parcel or a particle, Eq. (2.2)
$\mathbf{V} = U + \imath V$	Horizontal mean velocity in complex form, Sect. 3.1
$\mathbf{V}_g = U_g + \imath V_g$	Geostrophic wind velocity in complex form, Sect. 3.1
$\mathrm{v_s}$	Terminal velocity for particles, Eq. (5.104)
$w_*(z)$	Free-convection velocity scale in the surface layer, Eq. (3.35)
w_e	Entrainment velocity, Eq. (3.92)
$W_* = w_*(h)$	Global free-convection velocity scale, Eq. (3.83)
$\boldsymbol{x} = (x, y, z)$	Space coordinates; parcel/particle position, Eq. (2.1)
$\mathbf{X} = \langle uw \rangle + \imath \langle vw \rangle$	Turbulent vertical momentum flux in complex form, Sect. 3.1
z_d	Displacement height for the mean velocity profile, Eq. (3.33)
z_0	Roughness length, Eqs. (3.30), (3.31)
z_{0e}	Effective roughness length, Sect. 4.4
z_r	Reference height, Sect. 2.2.5.2
ε	Viscous dissipation of TKE, Eq. (2.74)
ε_θ	Viscous dissipation of temperature variance, Eq. (2.77)
ΔS	Speed-up, Eq. (4.24)
Δu	Perturbation velocity in the inner layer, Sect. 4.3.4; velocity difference for parcel pairs, Sect. 5.3
ΔU	Perturbation to the mean velocity profile, Eq. (4.16)
$\Delta \tau$	Perturbation to the shear stress, Eq. (4.17)
$\Delta \theta$	Temperature jump at the inversion level for CBL, Sect. 3.6.2
$\zeta = \|z/L\|$	Nondimensional vertical coordinate in the unstable surface layer, Sect. 3.4.2
$\zeta = z/\Lambda$	Nondimensional vertical coordinate in the stable boundary layer, Sect. 3.7
θ_*	Temperature scale in the surface layer, Eq. (3.18)
θ_{**}	Temperature scale in the free-convection sublayer, Eq. (3.36)

$\theta_f = \theta_a + \tilde{\theta}$ $= \theta_{00} + \theta_0 + \tilde{\theta}$ $= \theta_{00} + \theta_0 + \Theta + \theta$	Potential temperature of the air, in Kelvin degrees, Sect. 2.2.5.2
$\Theta_f = \theta_{00} + \theta_0 + \Theta$	Average temperature of the flow, Sect. 2.4.4
Θ_{**}	Temperature scale in the CBL, Eq. 3.88
κ	von Karman constant, reference value 0.4 , Sect. 3.3
κ_c	Molecular diffusivity of the scalar c [$m^2\ s^{-1}$], Sect. 2.2.4
κ_H	Thermal diffusivity, Sect. 2.2.6. For the air 2.1 10^{-5} m^2 s^{-1} at 20 C
Λ	Local value of the Obukhov length, Eq. (3.102)
ν	Kinematic viscosity of the fluid, Sect. 2.2.1. For the air at the standard temperature of 20 C and the pressure of 1000 hPa $\nu = 1.5\ 10^{-5}\ m^2\ s^{-1}$
$\rho_f = \rho_a + \tilde{\rho}$ $= \rho_{00} + \rho_0 + \tilde{\rho}$	Density of the air, Sect. 2.2.1
ρ_p	Inertial particle density, Sect. 5.7
τ	Time lag; $\tau = \|\mathbf{X}\|$: modulus of the turbulent vertical momentum flux, Sect. 3.1
τ_{ij}	Area averaged value of $-\langle u_i u_j \rangle$, Eq. (4.56)
Φ_U	Nondimensional vertical gradient of mean velocity, Eq. (3.38)
Φ_Θ	Nondimensional vertical gradient of mean potential temperature, Eq. (3.47)
σ	Square root of the variance of a stochastic variable, Eq. (2.32)

Chapter 1
Introduction

Abstract In the Introduction the planetary boundary layer (PBL) is described in general, as the part of the atmosphere where turbulence acts driving exchange processes and dispersion. Attention is paid to field and laboratory measurements, as well as to the use of numerical experiments as a further tool for knowledge.

1.1 The Basic Definition of the Planetary Boundary Layer

The planetary boundary layer (PBL) is the lower part of the troposphere, where the interactions with the surface of the Earth occur.

Similarly to all the boundary layers that develop as a fluid flows over a surface, the PBL is (under suitable conditions, which normally are verified for the atmosphere) characterized by the turbulence, that affects the exchange processes. For this reason, in this textbook we shall discuss about turbulence. The interaction with the surface occurs due to the exchange of momentum, of heat and of scalars (like the water vapour): the surface (the bare ground, a vegetative or a urban canopy, the sea) is a sink of momentum, but can be a source or a sink of heat or other scalars. Understanding these interactions is an important step for the proper modelling of weather and climate, and, in general, of the dynamics of the atmosphere.

The turbulence affects the transport of tracers (pollutants), which is a relevant issue in the air quality applications, and, more generally, in the study of the composition of the atmosphere (linked with climate). Besides the applications, the transport problem is related to the intimate nature of the turbulent flows, so that it deserves special attention also from a fundamental (theoretical) point of view.

Turbulent boundary layers are not limited to atmospheric flows: the general findings are relevant in geophysics as well as in engineering. However, some features are specific for the atmosphere, and will be detailed as possible.

F. Tampieri, *Turbulence and Dispersion in the Planetary Boundary Layer*,
Physics of Earth and Space Environments, DOI 10.1007/978-3-319-43604-3_1

1.2 A Few Words About Turbulence

An exhaustive treatment of this issue is well beyond the science of the author and the scope of this book. The reader must refer to the many textbooks starting from Monin and Yaglom (1971, 1975) for a classical introduction, with attention to geophysical applications, to the apparently simple Tennekes and Lumley (1972), to Pope (2000) in particular for turbulence modelling, to Wyngaard (2010) again focusing on the atmosphere. A suggested reading to go into the specific, but widely quoted, argument of the inertial subrange is Frisch (1995).

Here a short summary is presented, for the aspects of direct interest for the present study. A turbulent flow is characterized by random features of the state variables (velocity, temperature, or tracer concentration, for instance), by the existence of a wide range of scales of the motion (in terms of space and time), and by mixing properties.

The Reynolds number $R_e = \mathrm{UL}/\nu$ characterizes the flow of a viscous fluid. Here U is a velocity scale, for instance the average velocity, and L a length scale, for instance the depth of the fluid, or the width of the channel. The internal, molecular, friction is measured by the kinematic viscosity ν. If $R_e \sim 1$ the flow is laminar (and predictable). As R_e increases beyond 1000, say, the hydrodynamic instabilities make unpredictable some features of the flow. The velocity, for instance, is continuous (the fluid is viscous), but the accelerations can be quite large, and for many practical purposes the flow velocity can be considered a stochastic variable. The consequence is the need to give statistical descriptions of many phenomena characterizing the turbulent flow, and to refer to the probability density functions of the state variables. A time record of the velocity components and of the temperature in an atmospheric boundary layer is reported in Fig. 1.1, which qualitatively illustrates the random features of the variables characterizing the flow.

A second relevant feature of the turbulent flows is the presence of a range of time and space scales of the motion; in other words, the stochastic variables are not white noise, but are correlated in time and space (there is an underlying structure). This aspect can be qualitatively understood looking at the time patterns of Fig. 1.1: the high frequency fluctuations are superimposed to a fluctuating trend, characterized by longer time scales. The motion is organized in eddies, with spatial scales that range from those imposed by the boundaries down to scales small enough that their specific (computed from their length and velocity scales) Reynolds number is small, and viscosity dominates.

The internal structure of the turbulent flow produces important effects on the mixing. At small R_e the transport of scalars is described by the Fick law, i.e. the small scale molecular motion produces a large scale transport of the scalar in the direction of minus the gradient of the mean concentration (down-gradient transport): the paradigmatic case of diffusion occurs. At large R_e the eddies produce mixing, but the existence of a continuous range of scales inhibits the straightforward application of the diffusive, Fick law, approach. In a turbulent flow, the eddies may be as large as the scale of the gradient, and the transport may become non-Fickian.

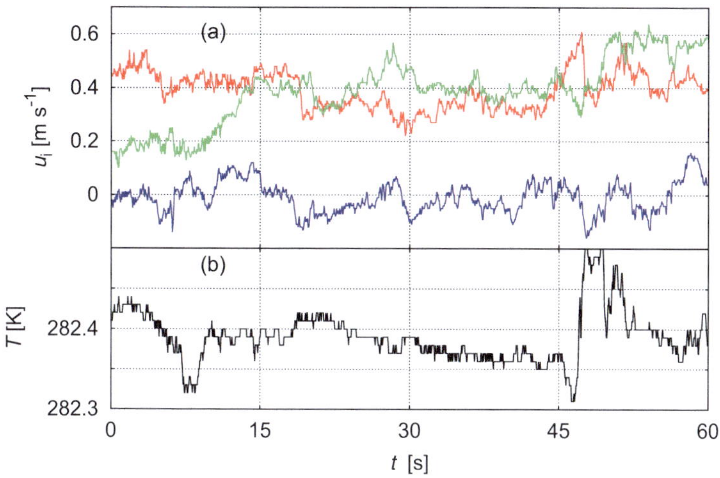

Fig. 1.1 Time series of the along-wind (*red line*), transversal (*green*) and vertical (*blue*) components of the air velocity fluctuations (**a**) and of the temperature (**b**), measured by a sonic anemometer at 7.5 m above the ground (CCT data). The high frequency record highlights the fluctuating behaviour of the observations and the different mean values of the three velocity components. Courtesy Mauro Mazzola, CNR ISAC

The space and time resolution of the observations and/or of the numerical models allow to resolve (i.e., explicitly describe) the larger scales of the motion. This aspect will be considered in detail in the following chapters; here it may be interesting to note that traditionally the motion of the atmosphere has been divided in mean wind and turbulence, just because the anemometers take an average over some time interval (of the order of minutes), so that the resolved variable (the averaged one) is the wind, the unresolved part is the turbulence. The arbitrariness is evident: from the point of view of seasonal dynamics of the atmosphere, the mid-latitude perturbations are eddies, i.e. turbulence.

1.3 The Structure and Evolution of the PBL

Quite schematically, the energy balance of the atmosphere in the PBL is related to the effect of the large scale motions of the atmosphere itself (winds and waves) and to the exchanges of momentum and heat at the surface. These exchanges are related to the radiation balance (incoming and outcoming radiation) and to the heat flux into the ground. Moisture effects can be relevant.

1.3.1 Local Equilibrium

As the vertical fluxes are large and the horizontal conditions pretty homogeneous, the PBL dynamics is dominated by the local conditions, i.e. its properties are only function of the distance from the surface, while the effects of the horizontal heterogeneities can be neglected. The local equilibrium is the basic paradigm for the study and the understanding of the phenomenology of the PBL, and will be discussed in Chap. 3.

Over a solid surface, if radiation is weak (overcast sky), the wind drives the turbulence and thus the exchanges. The turbulence is mainly produced by the shear, while thermal effects are minor: these are called quasi-neutral conditions (perfectly neutral conditions, occurring in absence of heat exchange, are probably realized only in the laboratory). In presence of a diurnal cycle, over the land, we shall take into account the time evolution of the radiative flux: the radiative flux during the day heats the ground and thus the air (turbulence increases, leading to the so-called unstable conditions) and cools both during the night (decreasing turbulence, leading to stable conditions). Intense heating of the ground gives origin to eddies of vertical size of the order of the PBL depth, with quite large positive vertical velocity: this is the onset of convection, a very efficient mechanism of mixing of the entire layer. Strong cooling damps the turbulence and the related exchanges, layers of air at different heights may be nearly independent on each other, the surface may become almost unimportant while phenomena occurring aloft have a relevant influence on the dynamics. The effect of radiation depends on the heat capacity of the surface, so that it is enhanced over the desert and almost negligible over the sea. The diurnal cycle disappears at high latitudes, leading to PBL characterized by a slow time evolution.

Remote sensing techniques based on the detection of density fluctuations and tracers in the atmosphere (SODAR and LIDAR respectively: see Sect. 1.5 for references to the instruments) allow the visualization of the vertical structure of the PBL: see Figs. 1.2, 1.3, 1.4 and 1.5.

Figures 1.2 and 1.3 depict the daily evolution of the PBL at midlatitudes, for a winter and a summer case, using a SODAR. During the night stable conditions prevail, and the density fluctuations (evidenced by the dark areas) appear to be highly variable in the vertical, and coherent in time, especially for the winter case (almost undetectable in the summer case). Note also the wave pattern at about 400 m from 3 h to 5 h on Feb. 8.

During the day the heating at the ground leads to the onset of convection, with large scale turbulent motions (of vertical size as the PBL itself) evidenced by the signals (dark lines) rapidly variable in time and vertically coherent. The convection starts around 6 h in the summer case and just before 10 h the plumes are so high to go beyond the vertical range of measure (i.e. the PBL depth in this case is greater than 800 m). In the winter case, convection begins around 8 h and is weaker: we can guess that the PBL depth reaches about 800 m for a few hours, around 14 h.

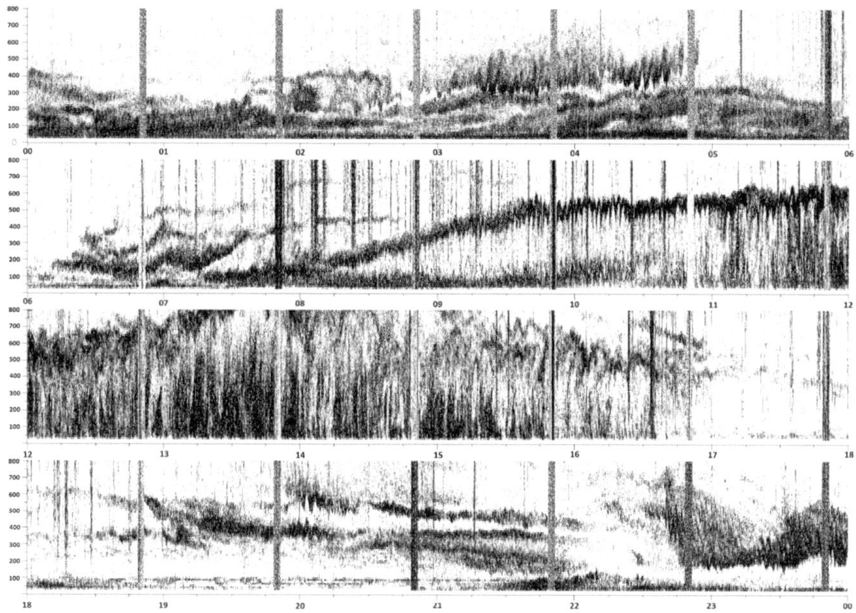

Fig. 1.2 SODAR echogram of the daily cycle in a winter day (February 08, 2011) at Castelporziano, a rural site near Roma, Italy. The local time (in hours) is reported in abscissa, the height (m above the ground) in ordinate. The *vertical gray stripes* correspond to calibration time intervals. Courtesy Angelo Viola, CNR ISAC, Roma

The development of a convective boundary layer is visualized also using LIDAR (which gives an estimate of the presence and the concentration of aerosol particles): see Fig. 1.4. Thanks to the larger measuring range, it shows the single convective cells that extends up to about 1.5 km in a typical summer sunny day.

A further SODAR sounding of the long lasting stable PBL in Antarctica is shown in Fig. 1.5. As expected, the absence of any time modulation contrasts with the mid-latitude cases; it is also worth noting the small depth of the layer, which reduces to a tenth of meters at the end of the period.

The presence of the diurnal cycle underlines the importance of the time evolution of the forcing mechanisms: the transitions from stable to unstable/convective conditions and vice versa, or the nocturnal radiative cooling, causing turbulence of decreasing intensity. Note that in general these phenomena occur on a time scale greater than the typical time scales of the turbulent flow, which then usually adjusts to the changing conditions.

Over the sea, some features characterize the PBL and distinguish it from the continental counterpart: the presence of a mobile lower boundary, which adjusts to a certain extent to the dynamics; the constant presence of moisture; the easier occurrence of homogeneous and steady conditions (apart from the coastal regions or in presence of cold/warm outbreaks); the diurnal cycle is small as well as the

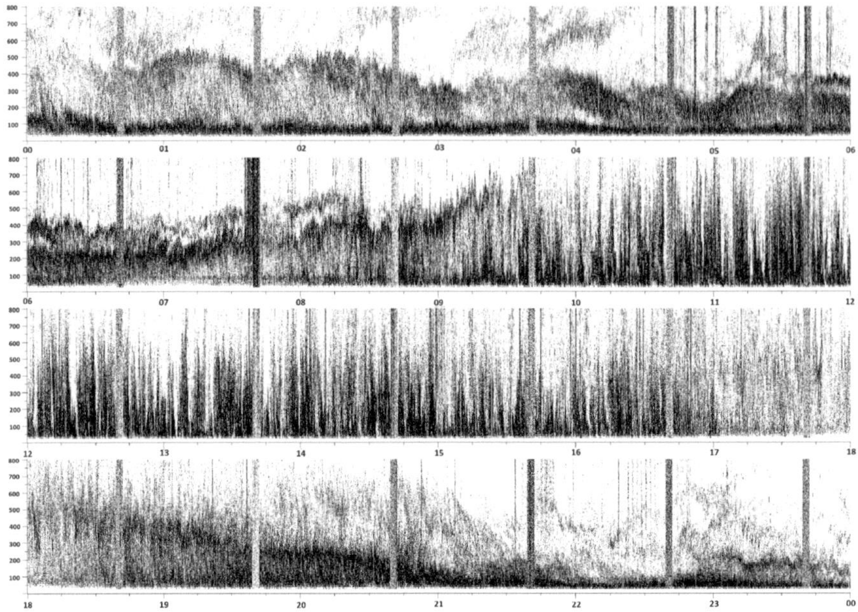

Fig. 1.3 As in Fig. 1.3, but for a summer day (August 31, 2010). Courtesy Angelo Viola, CNR ISAC, Roma

departures from near-neutral conditions (the air mass is in thermal equilibrium with the surface); large-scale eddies appears in form of rolls (organized motion).

1.3.2 Heterogeneities and Unsteadiness

In the real world, horizontal heterogeneities and unsteadiness on short time scales often occur, which means that the picture of the previous section must be revised, and the paradigm of local equilibrium loses its general validity, as discussed in detail in Chap. 4.

Broadly speaking, we can recognize two different situations in which heterogeneity is relevant, which we call ‘large scale’ and ‘small scale’, that undergo different treatments. The term ‘large scale’ means that we are able to make measurements and/or to realize numerical simulations and/or physical models which describe explicitly this situation. Its counterpart (the ‘small scale’ effects) occurs for heterogeneities/unsteadyness on space/time scales smaller than those we are able, or we want, to consider explicitly.

Large scale effects can be analyzed in detail; for instance, topographic features (hills and valleys, coastal borders) affect the wind field and the heating of the ground, thus produce horizontal variations of the forcing. Local circulations arise

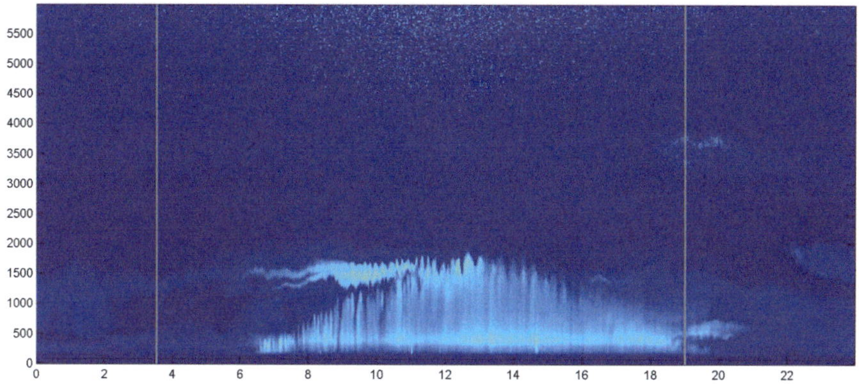

Fig. 1.4 LIDAR sounding of the convective boundary layer, for June 27, 2012, at San Pietro Capofiume, a site in the Po Valley, Italy. On the abscissa, time in hours, UTC. On the ordinate, the height in m above the ground. Courtesy Gianpaolo Gobbi, CNR ISAC, Roma. The presence of aerosols is evidenced by the almost *white areas*; *blue* means no particles. The *two vertical lines* correspond to sunrise and sunset. Further informations about the structure of the atmosphere on this day are reported in Sect. 3.6

(which are outside the scope of this book, but must be taken into account to understand the real, not too idealized, PBL). Changes of surface characteristics (grass to trees, land to sea) produce internal boundary layers; for instance, near the coastal line, during the night in the cold season, in presence of a land breeze, cold (stable) air blows over a warmer sea, generating an unstable internal PBL. The transitions related to the diurnal cycle (cited above) are a typical example of large scale unsteadiness.

Small scale effects are considered in terms of modifications of the rules characteristic of the local equilibrium paradigm: we abdicate the universality of the rules, maintaining some formal features and parameterizing the dependence on the small scale in the numerical value of the coefficients which appears in the formulas. The main consequence is the difficulty in finding general formulations of the parameterization.

The trend for tackling heterogeneous conditions is to refine the scale of the description, moving from parameterization to explicit treatment, thanks to the increasing computational power and the improvement of the observations. On the other hand, it is sometimes almost impossible, or unconvenient, to deal explicitly with all the details, while an averaged description may be all that we need. Situations like the wind and the vertical exchanges in a forest, or the meandering of the wind when the wind itself is quite low are examples of problems that typically do not require (or even do not permit) a detailed description. Note that both the 'large scale' and the 'small scale' approaches occur in the investigation on the urban PBL, and the different descriptions coexist in the common practice.

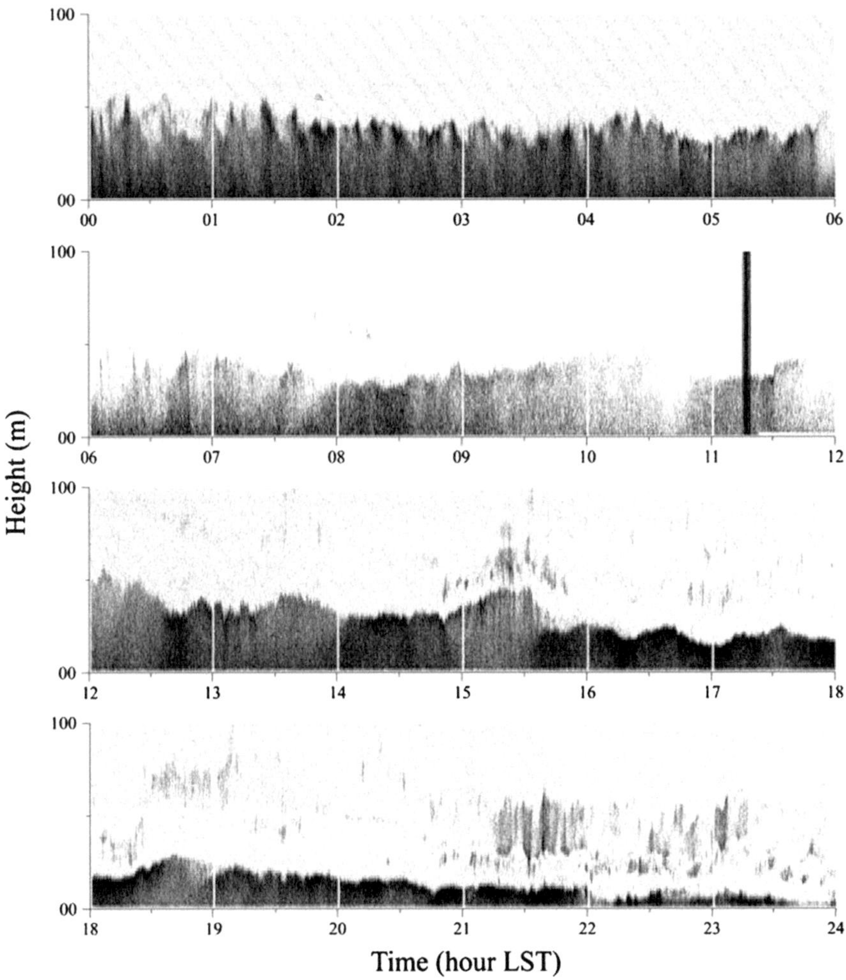

Fig. 1.5 Echogram of the stable boundary layer during the winter (August 20, 2012) at Concordia Station, Dome C, Antarctica, by an high resolution Surface-Layer MiniSODAR. Heights in m, local time in hours. Courtesy Stefania Argentini, CNR ISAC, Roma

1.3.3 The Boundary Layer Depth

The PBL depth h can be broadly defined as the height at which the interaction of the tropospheric flow with the surface becomes negligible. In spite of the fact that it is not a directly measurable quantity, a lot of words are spent about its determination, essentially because of the practical importance in modelling applications. This depth is quite evident in Fig. 1.4 as the level at which the aerosol sharply disappears, or in

Fig. 1.2 or Fig. 1.3 as the level at which convection stops, during daytime. Note thus that the depth is well identifiable in some conditions, and less well in others.

The traditional analysis of the PBL is largely based on the assumption that the surface fluxes play the major role in its dynamics and evolution, and h is the height at which these surface fluxes become negligible. The straightforward consequence is that the turbulent fluxes (at least of momentum and heat, but not only) are functions of the height z.

However, the surface fluxes may be small, or not relevant. In convective conditions, the mixing at the PBL top (i.e., the downwards heat flux from the troposphere) contributes to the growth of the layer. In stable conditions, the vertical fluxes may become negligible at the surface, while the turbulence is produced by the wind shear or by the waves aloft. In such cases the basic definition remains correct, but h cannot be estimated from surface fluxes.

Earth rotation introduces a further limitation to the PBL growth. So the final statement is that h is related to the surface fluxes, to the stability and wind conditions aloft (at its top), and to the Coriolis effect, if the geometry of the surface (hills and valleys) can be neglected.

1.4 The Transport Problem and the Turbulent Dispersion

The behaviour of a tracer advected by a turbulent flow shows complex, chaotic features. The tracer may be a pollutant, may be the temperature, may be a microorganism living in the sea. Some tracers are characterized by an almost infinite living time: they do not react with the environment; others combine by chemical reactions. Radioactive tracers change their properties according to their decay time.

Some tracers behave like the fluid molecules, i.e. their velocity is at any time equal to that of the flow: we shall refer to as fluid parcels. A fluid parcel is supposed to be an ideal small volume of fluid which can be identified: in general gaseous pollutants behave as parcels.

Other tracers are subject to the gravity acceleration and have their own dynamics, due to the inertia, like aerosols or water droplets: we shall refer to as inertial particles. In general, their velocity is different from that of the flow, and at least in certain conditions this aspect becomes critical.

Last but not least, some substances interact with the dynamics of the flow: for instance, temperature changes can occur associated with chemical reactions (this aspect will not be treated in this book).

The transport problem is tackled by computing the trajectories of the parcels, or of the particles, which means that their velocities must be known as function of time. In a turbulent flow, characterized by a wide range of scales of motion, this computation can be done in general for the large scales, but not for the small scales (a notable exception is given by the use of direct numerical simulations (DNS) of the flow). The problem is solved in a statistical sense, that is, in the computation of some moments of the positions of the parcels (or particles), or of their relative distance (in other words, in the determination of the mean concentration field, or of the variance field of the concentration fluctuations).

The diffusive paradigm (true for viscous flows) implies the existence of a diffusion coefficient, proportional to the product of a velocity scale and a length scale; the square of the size of a tracer cloud grows with time proportionally to the product of the diffusion coefficient and of the time. The diffusion concept is fundamentally incorrect as applied to turbulence; however it will be used as an interpretative tool, and also for parameterization purposes. We shall speak of dispersion in order to include all the complexities of the turbulent transport, and this will be treated in detail in Chap. 5.

1.5 Observations

Experiments for obtaining data on the PBL can be done in the field and in the laboratory.

The main advantage of the laboratory is the possibility to control the conditions of the experiment, and to repeat it. The main shortcoming is the difficulty in reaching what is called dynamical similarity: for instance, values of the Reynolds number large enough to reproduce atmospheric turbulence. We will refer, directly or indirectly, to laboratory observations made in wind tunnel and water: for instance, the measurements of high-order moments of the probability density function of velocity by Durst et al. (1987), the observations of dispersion in the convective boundary layer by Snyder et al. (2002), the velocity measurements made by particle tracking cited by Walpot et al. (2007) or by Toschi and Bodenschatz (2009).

The main advantage and at the same time the main shortcoming of the field observations is the opportunity to observe the real world, which changes in time and doesn't repeat itself: the impossibility to perform a true experiment in the Galileian sense is a big problem in the interpretation of the data and in the development of any theory. Field observations have been often arranged as specific campaigns, selecting site and climatological conditions to go as closer as possible to ideal conditions (to go closer to laboratory experiments).

Observations in the PBL have a long story (Lenschow 1986): the main instruments for probing turbulence are point measuring devices, like anemometers and thermometers, but also devices to measure concentration of tracers, and remote sensing devices, like SODAR and LIDAR.

Point measuring devices increased steadily the frequency of observation (both as far as velocity and scalars are concerned): sonic anemometers allow to determine the time series of the quantity under investigation with great detail (sampling frequency reach 100 Hz). Such devices need to be mounted on a tower, or a mast, which can induce perturbations in the flow from specific directions, or on aircrafts. Tower-mounted instruments allowed the beginning of the modern research on PBL: the Kansas experiment (Haugen et al. 1971), and the Wangara experiment (Hess et al. 1981) are two important examples. Some recent field experiments are cited in the

Appendix; the detailed description of each experiment (which can be found from the references) will allow the reader to appreciate the complexity of the setup, and the evolution of the instrumentation used.

Aircrafts have been since long time a platform for measurements; recent advancements of the knowledge of the PBL are related to the development of very fast sensors, which allow a good horizontal resolution in spite of the flight velocity (see, for instance, Malinowski et al. 2011).

Remote sensing instruments which use sound (SODAR: see Clifford et al. 1994) or electromagnetic (LIDAR: see Kovalev and Eichinger 2004) waves to investigate the flow features are unaffected by the problem of perturbate the flow, and their performances are increasing, but the spatial resolution is often a limitation.

1.6 Numerical Experiments and Simulations

As noted by Wyngaard (2004), the first numerical simulations of turbulent flows were proposed by Lilly (1962) more or less at the same time in which the turbulence community had the feeling of the enormous difficulties related to the analytical investigation, as focussed in the Marseille meeting held in 1961 (Moffatt 2002).

The capability to obtain numerical solutions of the governing equations has brought to use these simulations to investigate specific aspects of the flow dynamics: as stated by Aref (1986), the ambition was to provide the digital counterpart of the laboratory experiment (interestingly, this paper was presented at a congress commemorating G.I. Taylor, who gave important analytical contributions to the turbulence theory).

Historically, numerical experiments started with turbulence simulations in simplified conditions, and has been developed for atmospheric applications. Numerical models are used to study the processes (thus referring to idealized conditions) and to simulate real world (for instance, the meteorological models). Models based on the explicit solution of the flow mean variables and on the statistical treatment of the fluctuations are used besides models that explicitly solve the larger scales of the motion. The main shortcoming is related to the approximations made in order to solve the equations for cases of atmospheric relevance (see Chap. 6); the main advantage is the possibility to control and to change the conditions of the experiment.

Appendix

Here some data sets, often referred to in this book, are briefly described, and reference to the data sources are given (the data sets contain much more informations: here I report only those relative to the elaborations, and the figures, of the book).

The reader is invited to use the data, to go deeper into the problems related to their elaboration and thus to the understanding of turbulence in the PBL.

Each subsection is labelled using the same abbreviation/acronym used in the text and in the figures to identify the data source.

SGS 2002

This field experiment was designed to investigate the spatial scale of turbulence (see Higgins et al. 2007). The data are used in this book to investigate the features of some statistical moments of the velocity components (Barberis 2007).

CABAUW

A description of the site and the experimental setup can be found in
http://projects.knmi.nl/cabauw/insitu/index2.htm.

The Cabauw mast is located in the western part of the Netherlands (51.971°N, 4.927°E). Anemometers and thermometers are placed at 10, 20, 40, 80, 140 and 200 m. Temperature is measured also at 2 m. Sonic anemometers are placed at 3, 60, 100 and 180 m. For the data, acknowledgements are due to Fred Bosveld of the Royal Netherlands Meteorological Institute.

CCT

This data set is based on observations made on a tower, installed at the end of 2009 about 2 km NW off the village of Ny-Ålesund (78°55′N, 11°52′E, Svalbard Islands, Norway), in the framework of the so called Climate Change Tower Integrated Project (CCT-IP). The tower is 34 m height and is equipped with four thermo-hygrometers and four propeller anemometers at the heights of 2, 4.8, 10.3, and 33.4 m. At 3.7 m, 7.5 m and at 21 m, sets composed by a sonic anemometer and a fast hygrometer or a gas analyser are also installed.

The address to ask for the data is
http://www.isac.cnr.it/radiclim/CCTower/

Thanks are due to the Department of Earth System Science and Environmental Technologies of the Italian National Research Council (CNR) for the construction and the maintenance of the Amundsen-Nobile Climate Change Tower and for most of the instrumentation, and to the Korea Polar Research Institute (KOPRI) for the set of the fast-response instruments installed at the height of 21 m.

SABLES98

SABLES98 data were obtained at the CIBA site, located in the centre of an 800 km^2 plateau (Montes Torozos), which is 840 m above sea level and surrounded by fairly level grass plains in northern Spain. The site was specially designed for boundary-layer investigations and chosen to fulfill the requirement of horizontal homogeneity. A complete description of the field campaign can be found in Cuxart et al. (2000) and Yagüe et al. (2006).

Low response measurements (5 Hz) of wind speed are available at five levels between 3 and 100 m, temperature at 15 levels between 0.2 and 50 m, wind direction at 10, 20 and 100 m. Sonic anemometers (20 Hz sampling rate) are located at 32, 13.5 and 5.8 m.

Thanks are due to Dr. Joan Cuxart and Dr. Carlos Yagüe for the use of SABLES98 data, as well as to all the people participating in this field campaign, which was funded by the Spanish Government (project CLI97-0343 and action CLI98-1479E). Dr. Javier Pelaez is acknowledged for his technical support at CIBA during the field campaign, and Prof. J.L. Casanova, Director of the CIBA, for his kind help.

ARTIST-CBL

The data refer to measurements performed in the Arctic (Hartmann et al. 1999; Gryanik and Hartmann 2002) using aircraft-mounted instruments, in the frame of the project Arctic Radiation and Turbulence Interaction Study (ARTIST).

Acknowledgements are due to Jörg Hartmann, Alfred-Wegener-Institute für Polar- und Meeresforschung, DE, who supplied the data.

Other Relevant Data Sets

Two other data sets must be cited here, because of the importance in the recent investigations especially on stable boundary layers:

- CASES99: see Poulos et al. (2002). Measurements during October 1999, in southeast Kansas; the main tower was 60 m high, with six sonic anemometers and slow response instruments placed alternatively at 5 m intervals.
- SHEBA: see Uttal et al. (2002); for a recent review of the data, see Grachev et al. (2013). One year of continuous data on the Arctic Ocean ice pack; a main tower with five measuring levels ($z = 2.2$, 3.2, 5.1, 8.9 and 18.2 m above the surface).

References

H. Aref, The numerical experiment in fluid mechanics. J. Fluid Mech. **173**, 15–41 (1986)

E. Barberis, Analisi statistiche nello strato limite turbolento, thesis. Univ. Torino, Dip. Fisica (2007)

S.F. Clifford, J.C. Kaimal, R.J. Lataitis, R.G. Strauch, Ground-based remote profiling in the atmospheric studies: an overview. Proc. IEEE **82**, 313–355 (1994)

J. Cuxart, C. Yagüe, G. Morales, E. Terradellas, J. Orbe, J. Calvo, A. Fernandez, M.R. Soler, C. Infante, P. Buenestado, A. Espinalt, H.E. Joergensen, J.M. Rees, J. Vila, J.M. Redondo, I.R. Cantalapiedra, L. Conangla, Stable atmospheric boundary-layer experiment in Spain (SABLES 98): a report. Bound.-Layer Meteorol. **96**, 337–370 (2000)

F. Durst, J. Jovanovic, L.J. Kanevce, Probability density distribution in turbulent wall boundary-layer flows, in *Turbulent Shear Flows*, vol. 5, ed. by F. Durst, B.E. Launder, J.L. Lumley, F.W. Schmidt, J.H. Whitelaw (Springer, Berlin, 1987)

U. Frisch, *Turbulence* (Cambridge University Press, Cambridge, 1995), 296 p.

A.A. Grachev, E.L. Andreas, C.W. Fairall, P.S. Guest, P.O.G. Persson, The critical Richardson number and limits of applicability of local similarity theory in the stable boundary layer. Bound.-Layer Meteorol. **147**, 51–82 (2013)

V. Gryanik, J. Hartmann, A turbulence closure for the convective boundary layer based on a two-scale mass-flux approach. J. Atmos. Sci. **59**, 2729–2744 (2002)

J. Hartmann et al., Arctic radiation and turbulence interaction study, Polar Res. Report. 305. Alfred Wegener Institute for Polar and Marine Sciences (1999)

D.A. Haugen, J.C. Kaimal, E.F. Bradley, An experimental study of the Reynolds stress and heat flux in the atmospheric surface layer. Q. J. R. Meteorol. Soc. **97**, 168–180 (1971)

G.D. Hess, B.B. Hicks, T. Yamada, The impact of the Wangara experiment. Bound.-Layer Meteorol. **20**, 135–174 (1981)

C.W. Higgins, C. Meneveau, M. Parlange, The effect of filter dimension on the subgrid-scale stress, heat flux, and tensor alignements in the atmospheric surface layer. J. Atmos. Ocean. Technol. **24**, 360–375 (2007)

V.A. Kovalev, W.E. Eichinger, *Elastic Lidar: Theory, Practice, and Analysis Methods* (Wiley, New York, 2004)

D.H. Lenschow (ed.), *Probing the Atmospheric Boundary Layer* (American Meteorological Society, Boston, 1986)

D.K. Lilly, On the numerical simulation of buoyant convection. Tellus **14**, 148–172 (1962)

S.P. Malinowski, K.E. Haman, M.K. Kopec, W. Kumala, H. Gerber, Small-scale turbulent mixing at stratocumulus top observed by means of high resolution airborne temperature and LWC measurements. J. Phys. Conf. Ser. **318**, 072013.1–072013.7 (2011)

H.K. Moffatt, G.K. Batchelor and the homogenization of turbulence. Ann. Rev. Fluid Mech. **34**, 19–35 (2002)

A.S. Monin, A.M. Yaglom, *Statistical Fluid Mechanics* vol. I (MIT Press, Cambridge, 1971), 769 pp.

A.S. Monin, A.M. Yaglom, *Statistical Fluid Mechanics*, vol. II (MIT Press, Cambridge, 1975), 874 pp.

S. Pope, *Turbulent Flows* (Cambridge University Press, Cambridge, 2000)

G.S. Poulos, W. Blumen, D.C. Fritts, J.K. Lundquist, J. Sun, S.P. Burns, C. Nappo, R.M. Banta, R.K. Newsom, J. Cuxart, E. Terradellas, B.B. Balsley, M. Jensen, CASES-99: a comprehensive investigation of the stable nocturnal boundary layer. Bull. Am. Meteorol. Soc. **83**, 555–581 (2002)

W.H. Snyder, R.E. Lawson Jr., M.S. Shipman, J. Lu, Fluid modelling of atmospheric dispersion in the convective boundary layer. Bound.-Layer Meteorol. **102**, 335–366 (2002)

H. Tennekes, J.L. Lumley, *A First Course in Turbulence* (MIT Press, Cambridge, 1972)

F. Toschi, E. Bodenschatz, Lagrangian properties of particles in turbulence. Ann. Rev. Fluid Mech. **41**, 375–404 (2009)

T. Uttal et al., Surface heat budget of the arctic ocean. Bull. Am. Meteorol. Soc. **83**, 255–275 (2002)

R.J.E. Walpot, C.W.M. van der Geld, J.G.M. Kuerten, Determination of the coefficients of Langevin models for inhomogeneous turbulent flows by three-dimensional particle tracking velocimetry and direct numerical simulation. Phys. Fluids **19**, 0451021–04510213 (2007)

J.C. Wyngaard, Changing the face of small-scale meteorology, in *Atmospheric Turbulence and Mesoscale Meteorology*, ed. by E. Fedorovich, R. Rotunno, B. Stevens (Cambridge University Press, Cambridge, 2004), pp. 17–34

J.C. Wyngaard, *Turbulence in the Atmosphere* (Cambridge University Press, Cambridge, 2010)

C. Yagüe, S. Viana, G. Maqueda, J.M. Redondo, Influence of stability on the flux-profile relationships for wind speed, ϕ_m, and temperature, ϕ_h, for the stable atmospheric boundary layer. Nonlinear Process. Geophys. **13**, 185–203 (2006)

Chapter 2
A Summary of Mathematics and Physics for PBL

Abstract This chapter is a summary of the mathematics and physics needed for understanding and modelling the PBL and the dispersion. Definition and equations are necessary for setting the nomenclature and the symbols. Probability density functions, correlation functions and spectra are introduced. References will be given for the derivations of the Navier-Stokes and thermodynamics equations as well as the equations for the averaged quantities. The inertial subrange theory and the eddy diffusivity concept will be presented here, with proper references for further readings.

2.1 Eulerian and Lagrangian Description

In this investigation about turbulence and dispersion in the planetary boundary layer we shall refer to two descriptions.

The Eulerian view is based on the equations describing the time evolution and space changes of the quantities characterizing the flow: velocity, temperature, concentration of tracers. For instance, the velocity field $\tilde{\boldsymbol{u}}(\boldsymbol{x}, t)$ is a function of position $\boldsymbol{x}$ and time t. Fixed point instruments supply Eulerian measurements.

The Lagrangian view aims to describe the evolution of the properties of a fluid parcel along its trajectory. The trajectory of a fluid parcel is given by:

$$\boldsymbol{x}(t) = \boldsymbol{x}(t_0) + \int_{t_0}^{t} \boldsymbol{v}(\tau) d\tau \ , \tag{2.1}$$

where the parcel velocity $\boldsymbol{v}(\tau)$ is given by the (Eulerian) velocity of the flow at $\boldsymbol{x}(\tau)$:

$$\boldsymbol{v}(\tau) = \tilde{\boldsymbol{u}}[\boldsymbol{x}(\tau), \tau] \ . \tag{2.2}$$

Note that (2.2) is not verified for a solid (inertial) particle; this question will be addressed when dealing with dispersion problems (Chap. 5).

F. Tampieri, *Turbulence and Dispersion in the Planetary Boundary Layer*,
Physics of Earth and Space Environments, DOI 10.1007/978-3-319-43604-3_2

2.2 The Equations for Velocity and Passive Scalars

The Navier-Stokes equations, the continuity equation and the conservation equation for a passive scalar are summarized here. Detailed derivation and discussion can be found in many textbooks, see, for instance, Monin and Yaglom (1971), or Wyngaard (2010), among many others.

2.2.1 *The Navier-Stokes Equations (NSE) in a Rotating Reference Frame*

The second law of dynamics written for the components of the fluid velocity vector $\tilde{\boldsymbol{u}} = (\tilde{u}, \tilde{v}, \tilde{w}) = (\tilde{u}_1, \tilde{u}_2, \tilde{u}_3)$ reads:

$$\frac{\mathrm{D}\tilde{u}_i}{\mathrm{D}t} = -\frac{1}{\rho_\mathrm{f}}\frac{\partial p_\mathrm{f}}{\partial x_i} + \varepsilon_{ij3} f \tilde{u}_j + \nu \frac{\partial^2 \tilde{u}_i}{\partial x_j \partial x_j} - \delta_{i3} g \,, \tag{2.3}$$

where

$$\frac{\mathrm{D}}{\mathrm{D}t} \equiv \frac{\partial}{\partial t} + \tilde{u}_j \frac{\partial}{\partial x_j} \tag{2.4}$$

is the material derivative (the change of the quantity under consideration following the fluid motion), p_f is the pressure and ρ_f is the density of the fluid, f is the Coriolis parameter (equal to two times the local angular velocity of the reference system: in a site with latitude ϕ, being Ω the angular velocity of the Earth, $f = 2\Omega \sin\phi$), ν is the kinematic viscosity of the fluid and g is the gravity acceleration. Note that we shall use both the notations for the velocity components, consistently with x, y, z or x_i, $i = 1, 2, 3$ for the spatial coordinates. As appropriate, the vector $\boldsymbol{v} = (v_1, v_2, v_3)$ will be used also to identify the velocity along a Lagrangian trajectory: the presence of the subscript and the context will eliminate risks of confusion with the Eulerian velocity component $v \equiv u_2$. Remember that $\delta_{ij} = 1$ for $i = j$ and $= 0$ for $i \neq j$; $\varepsilon_{ijk} = 1$ for $ijk = 123, 231, 312, = -1$ for $ijk = 321, 213, 132$ and $= 0$ otherwise.

In the present study, it is convenient to split the variables into a contribution related to the ambient at rest, denoted by the subscript a, and a contribution related to the fluid motion (the dynamics), denoted by a tilde: thus the pressure is written $p_\mathrm{f} = p_\mathrm{a} + \tilde{p}$ and the density $\rho_\mathrm{f} = \rho_\mathrm{a} + \tilde{\rho}$.

2.2.2 The Hydrostatic Pressure and the PBL Form of NSE

In the study of the PBL, the large-scale dynamics of the atmosphere may be considered a basic state, and (2.3) is rewritten accordingly.

Let us consider the air as a perfect gas, so that:

$$p_\mathrm{f} = R\rho_\mathrm{f} T_\mathrm{f} \,, \tag{2.5}$$

where $T_\mathrm{f} = T_\mathrm{a} + \tilde{T}$ is the absolute temperature and R is the gas constant for dry air.

At rest, i.e. $\tilde{\boldsymbol{u}} = 0$, the values of state variables of the fluid reduce to the ambient ones: $\rho_\mathrm{f} = \rho_\mathrm{a}$, $p_\mathrm{f} = p_\mathrm{a}$ and $T_\mathrm{f} = T_\mathrm{a}$. Thus (2.5) holds for the ambient variables, and (2.3) reduces to:

$$\frac{\mathrm{d}p_\mathrm{a}}{\mathrm{d}z} = -\rho_\mathrm{a} g = -\frac{g p_\mathrm{a}}{R T_\mathrm{a}} \,, \tag{2.6}$$

which allows to compute the hydrostatic pressure p_a (the pressure of the fluid at rest) as a function of height, given the temperature profile.

Subtracting (2.6) from (2.3) the equation for the velocity components reads:

$$\frac{\mathrm{D}\tilde{u}_i}{\mathrm{D}t} = -\frac{1}{\rho_\mathrm{f}}\frac{\partial \tilde{p}}{\partial x_i} + \varepsilon_{ij3} f \tilde{u}_j + \nu \frac{\partial^2 \tilde{u}_i}{\partial x_j \partial x_j} - \delta_{i3} g \frac{\tilde{\rho}}{\rho_\mathrm{a}} \,. \tag{2.7}$$

We assume that the dynamics produces small departures from the hydrostatic pressure. Thus (see Exercise 2.1):

$$\frac{\tilde{\rho}}{\rho_\mathrm{a}} \approx -\frac{\tilde{T}}{T_\mathrm{a}} \,, \tag{2.8}$$

so that (2.7) can be written:

$$\frac{\mathrm{D}\tilde{u}_i}{\mathrm{D}t} = -\frac{1}{\rho_\mathrm{f}}\frac{\partial \tilde{p}}{\partial x_i} + \varepsilon_{ij3} f \tilde{u}_j + \nu \frac{\partial^2 \tilde{u}_i}{\partial x_j \partial x_j} + \delta_{i3} g \frac{\tilde{T}}{T_\mathrm{a}} \,. \tag{2.9}$$

Equations (2.7) and (2.9) may be further approximated writing the ambient density as a constant part plus a contribution changing with height and possibly with time: $\rho_\mathrm{a} = \rho_{00} + \rho_0(z,t)$, with $\partial\rho_0/\partial z \ll \rho_{00}/\mathrm{H}$, being H the height scale, for instance, the depth of the PBL. Hereinafter we shall substitute ρ_f and ρ_a with ρ_{00} in (2.7) and (2.9).

These equations highlight that the PBL dynamics is related to the departures from the basic state of the atmosphere, characterized by the pressure vertical profile (in hydrostatic equilibrium) and by the related profiles of temperature and density.

The order of magnitude of the dynamic contribution to the pressure can be evaluated for a steady flow in absence of external forces and internal friction, by comparing the advection and pressure terms in (2.7). For one component it results $\rho_{00}\tilde{u}\,\mathrm{d}\tilde{u}/\,\mathrm{d}x \approx -\mathrm{d}\tilde{p}/\,\mathrm{d}x$ which can be written $O(\tilde{p}) \sim \rho_{00}O(\tilde{u}^2)$.

2.2.3 *The Continuity Equation*

The continuity equation expresses the mass conservation: the time variation of the mass in a fixed volume is equal to minus the divergence of the flux through the volume boundaries:

$$\frac{\partial \rho_{\mathrm{f}}}{\partial t} + \frac{\partial\left(\tilde{u}_i \rho_{\mathrm{f}}\right)}{\partial x_i} = 0\,. \tag{2.10}$$

For a constant density fluid, (2.10) becomes:

$$\frac{\partial \tilde{u}_i}{\partial x_i} = 0\,. \tag{2.11}$$

In this book we shall use the Boussinesq approximation (see, for instance, Spiegel and Veronis 1960), i.e. we shall neglect variations of the thermodynamic properties of the fluid, but as far the buoyancy force (Archimedes force) is concerned.

2.2.4 *The Equation for a Passive Scalar*

The equation for the concentration $\tilde{c}$ of a passive scalar (in units of mass over volume), in absence of sources or sinks, derives from the conservation relation, analogous to (2.10):

$$\frac{\partial \tilde{c}}{\partial t} = -\frac{\partial F_i^{(\mathrm{c})}}{\partial x_i}\,, \tag{2.12}$$

where $\boldsymbol{F}^{(\mathrm{c})}$ is the flux of $\tilde{c}$, which is given by an advective term, due to the fluid motion, and a diffusive term, due to the presence of spatial gradients of the scalar. The last term is expressed by the empirical rule, called Fick law (Csanady 1973, p. 4), which puts in relation the diffusive flow of $\tilde{c}$ with its gradient, via the molecular diffusivity κ_{c} specific for the scalar under consideration. Thus:

$$F_i^{(\mathrm{c})} = \tilde{u}_i \tilde{c} - \kappa_{\mathrm{c}} \frac{\partial \tilde{c}}{\partial x_i}\,, \tag{2.13}$$

and (2.12) reads

$$\frac{\partial \tilde{c}}{\partial t} = -\frac{\partial}{\partial x_i}\left(\tilde{u}_i\tilde{c} - \kappa_c\frac{\partial \tilde{c}}{\partial x_i}\right) . \tag{2.14}$$

Equation (2.14) may be rewritten as:

$$\frac{D\tilde{c}}{Dt} = -\tilde{c}\frac{\partial \tilde{u}_i}{\partial x_i} + \kappa_c\frac{\partial^2 \tilde{c}}{\partial x_i\,\partial x_i} , \tag{2.15}$$

which shows that in a constant density fluid, the concentration of a passive scalar is conserved, if the molecular diffusion is negligible.

In a compressible fluid, even if the molecular diffusion can be neglected, the concentration is not conserved because $\partial\tilde{u}_i/\partial x_i \neq 0$. From (2.15), the equation for the mixing ratio $\tilde{q}_c = \tilde{c}/\rho_f$ (mass of the scalar over total mass of the fluid) can be derived (see Exercise 2.2):

$$\frac{D\tilde{q}_c}{Dt} = \frac{\kappa_c}{\rho_f}\frac{\partial^2 \tilde{c}}{\partial x_i\,\partial x_i} , \tag{2.16}$$

which shows that the mixing ratio $\tilde{q}_c$ is conserved in a compressible fluid, if molecular diffusion can be neglected.

2.2.5 A Little Thermodynamics

For a detailed presentation of this argument, see for instance Gill (1982). Let consider the thermodynamics of a parcel of air: E is the internal energy per unit mass, Q the heat content and W the work made during a reversible transformation. The work is expressed by the change in the specific volume $\upsilon_s = \rho_f^{-1}$ at constant pressure $dW = -p_f\, d\upsilon_s$. Then:

$$dE = dQ - p_f\, d\upsilon_s . \tag{2.17}$$

For an adiabatic transformation $dQ = 0$.

Let consider the entropy per unit mass S, so that $dQ = T\,dS$ (i.e., a reversible adiabatic transformation is isentropic). For a reversible transformation (2.17) reads:

$$T_f\, dS = c_p\, dT_f - \rho_f^{-1}\, dp_f , \tag{2.18}$$

where c_p is the specific heat at constant pressure.

The previous result can be applied to an air parcel moving adiabatically (i.e. with $\mathrm{d}S = 0$) along a trajectory. The equation for the temperature results:

$$\rho_\mathrm{f} c_\mathrm{p} \frac{\mathrm{D}T_\mathrm{f}}{\mathrm{D}t} - \frac{\mathrm{D}p_\mathrm{f}}{\mathrm{D}t} = 0 \,. \tag{2.19}$$

2.2.5.1 The Adiabatic Gradient of Temperature

The temperature change occurring during the adiabatic vertical motion of a fluid parcel is computed by substituting the material derivative in (2.19) with the expression $\tilde{w}\,\mathrm{d}/\,\mathrm{d}z$ and using (2.6). Thus the temperature equation reduces to:

$$\frac{\mathrm{d}T_\mathrm{a}}{\mathrm{d}z} = -\frac{g}{c_\mathrm{p}} \,. \tag{2.20}$$

With typical values for dry air it results $\mathrm{d}T_\mathrm{a}/\,\mathrm{d}z \approx -10\,\mathrm{K\,km^{-1}}$: this is the equilibrium vertical gradient of the temperature in absence of horizontal motion and in steady conditions.

2.2.5.2 The Potential Temperature

The potential temperature is the temperature that a fluid parcel with temperature T and pressure p at an height z takes if it moves adiabatically to a reference height z_r with pressure $p_\mathrm{f}(z_\mathrm{r})$. In other words, the potential temperature is constant if the entropy per unit mass S is constant.

With reference to (2.18), the potential temperature of the fluid θ_f is defined as the quantity that satisfies the equation:

$$\frac{c_\mathrm{p}}{T_\mathrm{f}}\frac{\mathrm{d}T_\mathrm{f}}{\mathrm{d}z} - \frac{1}{\rho_\mathrm{f} T_\mathrm{f}}\frac{\mathrm{d}p_\mathrm{f}}{\mathrm{d}z} = \frac{c_\mathrm{p}}{\theta_\mathrm{f}}\frac{\mathrm{d}\theta_\mathrm{f}}{\mathrm{d}z} \,. \tag{2.21}$$

Integrating from the reference height z_r to z and noting that by definition $T_\mathrm{f}(z_\mathrm{r}) = \theta_\mathrm{f}(z_\mathrm{r})$, it results:

$$\theta_\mathrm{f}(z) = T_\mathrm{f}(z)\left[\frac{p_\mathrm{f}(z)}{p_\mathrm{f}(z_\mathrm{r})}\right]^{-R/c_\mathrm{p}} \,. \tag{2.22}$$

A common choice is to use the surface value of the pressure as reference: $p_\mathrm{f}(z_\mathrm{r} = 0)$. For atmospheric conditions, $R/c_\mathrm{p} \approx 0.286$. The potential temperature is splitted as follows: $\theta_\mathrm{f} = \theta_\mathrm{a} + \tilde{\theta} = \theta_{00} + \theta_0(z,t) + \tilde{\theta}(x,y,z,t)$. The potential temperature is constant in an adiabatic atmosphere (see Exercise 2.3).

2.2.5.3 The Virtual Temperature

In order to account for the effect of the water vapour on the air density, the virtual temperature is defined as follows.

The fluid pressure is splitted in the partial pressure of the dry air and that of water vapour $p_\mathrm{f} = p_\mathrm{d} + p_\mathrm{v}$. For dry air (of molecular mass 29) $p_\mathrm{d} = \rho_\mathrm{d}\, R\, T_\mathrm{f}$ and for the water vapour (of molecular mass 18) $p_\mathrm{v} = \rho_\mathrm{v}\, R_\mathrm{v}\, T_\mathrm{f}$. For the mixture, the specific humidity is defined as $q_\mathrm{f} = \rho_\mathrm{v}/\rho_\mathrm{f}$. So $\rho_\mathrm{f} = \rho_\mathrm{d} + \rho_\mathrm{v}$ and $\rho_\mathrm{d} = (1 - q_\mathrm{f})\rho_\mathrm{f}$.

By taking the ratio of the perfect gas equations for vapour and dry air it results:

$$\frac{p_\mathrm{v}}{p_\mathrm{f}} = \frac{q_\mathrm{f}}{R/R_\mathrm{v} + (1 - R/R_\mathrm{v})q_\mathrm{f}} , \tag{2.23}$$

and from the equation of state:

$$\rho_\mathrm{f} \equiv \frac{p_\mathrm{d}}{RT_\mathrm{f}} + \frac{p_\mathrm{v}}{R_\mathrm{v}T_\mathrm{f}} = \frac{p_\mathrm{f}}{RT_\mathrm{f}[1 - q_\mathrm{f} + q_\mathrm{f}/(R/R_\mathrm{v})]} . \tag{2.24}$$

This equation suggests to define the virtual temperature as $T_\mathrm{v} = T_\mathrm{f}(1 - q_\mathrm{f} + q_\mathrm{f}/(R/R_\mathrm{v})) \approx T_\mathrm{f}(1 + 0.608\, q_\mathrm{f})$. The definition of the potential virtual temperature ϑ_v is straightforward.

2.2.6 *The Equations for the Temperature and for the Potential Temperature*

When the motion is non isentropic (non adiabatic), additional terms must be included in (2.19). The radiative exchange is quantified via the divergence of the radiative flux density $\boldsymbol{F}^{(\mathrm{r})}$ (energy exchanged with the ambient air). The heat flux in direction i due to the molecular conduction is proportional to the temperature gradient, and is given by $-(\rho_{00}\kappa_\mathrm{H}c_\mathrm{p})\partial T_\mathrm{f}/\partial x_i$ where $\rho_{00}\kappa_\mathrm{H}c_\mathrm{p}$ is called the thermal conductivity and κ_H the thermal diffusivity of the air. The heating due to the changes of phase (latent heat release), to the chemical reactions and to the viscous dissipation, is represented by the rate of heating for unit volume Q_H.

Thus, (2.19) is modified to take into account the additional effects and reads:

$$\rho_\mathrm{f}c_\mathrm{p}\frac{\mathrm{D}T_\mathrm{f}}{\mathrm{D}t} - \frac{\mathrm{D}p_\mathrm{f}}{\mathrm{D}t} = \frac{\partial}{\partial x_i}\left[(\rho_{00}\kappa_\mathrm{H}c_\mathrm{p})\frac{\partial T_\mathrm{f}}{\partial x_i} - F_i^{(\mathrm{r})}\right] + Q_\mathrm{H} . \tag{2.25}$$

Using (2.21) for a generic motion, i.e. substituting $\mathrm{d}/\,\mathrm{d}z$ with $\mathrm{D}/\mathrm{D}t$, the equation for the potential temperature results:

$$\rho_\mathrm{f}c_\mathrm{p}T_\mathrm{f}\theta_\mathrm{f}^{-1}\frac{\mathrm{D}\theta_\mathrm{f}}{\mathrm{D}t} = \frac{\partial}{\partial x_i}\left[(\rho_{00}\kappa_\mathrm{H}c_\mathrm{p})\frac{\partial T_\mathrm{f}}{\partial x_i} - F_i^{(\mathrm{r})}\right] + Q_\mathrm{H} . \tag{2.26}$$

Neglecting density variations, i.e. referring to a constant density fluid, assuming the ratio T_f/θ_f approximately constant (see Wyngaard 2010, p. 180) and neglecting the dependence of θ_0 on time, the equation for the potential temperature, in absence of radiative exchanges, chemical effects and dissipation, results:

$$\frac{D\tilde{\theta}}{Dt} = \tilde{w}\frac{d\theta_0}{dz} + \frac{\partial}{\partial x_i}\left(\kappa_H \frac{\partial\tilde{\theta}}{\partial x_i}\right) . \tag{2.27}$$

Hereinafter we will always use the potential temperature (and call it shortly temperature), unless explicitly stated. To be consistent we will use in (2.9) $\tilde{\theta}/\theta_{00}$ in place of $\tilde{T}/T_a$. The virtual potential temperature is used in the treatment of the observations, to account for the presence of water vapour, in absence of changes of state.

2.2.7 The Nondimensional Form of the Equations

Equations (2.7) and (2.27) can be written in nondimensional form using a velocity scale U and a length scale L. Obviously, the time scale turns out to be L/U. The scale for the pressure is ρ_{00} U^2; the scale for the density is $L\partial\rho_0/\partial z$.

From (2.7) it results:

$$\frac{D\tilde{u}_i}{Dt} = -\frac{\partial\tilde{p}}{\partial x_i} + \varepsilon_{ij3}R_o^{-1}\tilde{u}_j + R_e^{-1}\frac{\partial^2\tilde{u}_i}{\partial x_j\partial x_j} + \delta_{i3}F_r^{-2}\tilde{\rho} , \tag{2.28}$$

where all the variables are nondimensional. The motion is characterized by the Rossby number $R_o = U/fL$, the Reynolds number $R_e = UL/\nu$ and the Froude number $F_r = U/NL$, that give an estimate of the importance respectively of the rotation (the Coriolis effect), the molecular viscosity and the static stability, with respect to the advection and pressure terms, assumed to be in balance. Note that the Brunt-Väisälä frequency of the basic state

$$N = (-g/\rho_{00}\,\partial\rho_0/\partial z)^{1/2} = (g/\theta_{00}\,\partial\theta_0/\partial z)^{1/2} , \tag{2.29}$$

which appears in the definition of the Froude number, is a measure of the atmospheric stability: N is meaningful (real) if the density decreases with height, or the temperature increases with height.

Similarly, (2.27) is made nondimensional using $L\partial\theta_0/\partial z$ as a scale for $\tilde{\theta}$. It results:

$$\frac{D\tilde{\theta}}{Dt} = \tilde{w} + P_r^{-1}R_e^{-1}\frac{\partial^2\tilde{\theta}}{\partial x_j\partial x_j} , \tag{2.30}$$

where $P_r = \nu/\kappa_H$ is the Prandtl number.

For this analysis we assume that a unique scale for the length and a unique scale for the velocity characterize the motion. It will be evident in the study of PBL that this assumption may be wrong in many cases, and more detailed analyses may be necessary. Two examples may be anticipated. First: the presence of a vertical gradient of density induces a vertical scale, which is independent on (and possibly quite smaller than) the horizontal one: when stratification is a relevant effect, it is necessary to consider different scales for the horizontal and vertical directions. Second: the structure of the turbulence, qualitatively described in the Introduction, is made by a continuous distribution of eddies of different sizes, from the larger ones, that we could imagine to be of scale L, to the smaller ones, that we will show later to be related to the viscosity of the fluid. Thus, in the same flow, the Reynolds number may be large for the larger scales (and the viscosity can be neglected), but it becomes small for the smaller scales (and viscosity is important).

2.3 Stochastic Variables

Although we known the differential equations governing the fluid motion, their solution for turbulent conditions is not known in general. In this book the variables that characterize the turbulence dynamics will be assumed to be stochastic variables (for a discussion, see Frisch 1995, Chap. 3), and we will derive and use the equations describing some statistical moments (the first-order and the second-order moments).

2.3.1 *Probability Density Function and Moments*

The description of the fluid motion in terms of stochastic variables entails the knowledge of the probability distribution of each variable, and the joint distribution of different variables. The atmospheric (in general, geophysical) observations allow to determine a few statistical moments of the variables, for instance of the velocity components, or of the temperature, and to guess, if necessary, the shape of the probability distribution.

For a detailed introduction to the topic of stochastic variables and calculus, see, for instance, Gardiner (1990).

Let define $F_\alpha(a)$ the probability of a stochastic scalar variable α to be smaller than a given value a. Then the probability density function is defined as $f_\alpha(a) = \mathrm{d}F_\alpha/\,\mathrm{d}a$ and allows to compute the moments of order n of the stochastic variable defined in the range (a_1, a_2):

$$\langle a^n \rangle = \int_{a_1}^{a_2} a^n f_\alpha(a)\,\mathrm{d}a\,, \tag{2.31}$$

where $n = 0$ gives the normalization condition, $n = 1$ the mean value, and so on.

The variance reads:

$$\sigma^2 = \int_{a_1}^{a_2} (a - \langle a \rangle)^2 f_\alpha(a)\, \mathrm{d}a = \langle a^2 \rangle - \langle a \rangle^2 \,, \tag{2.32}$$

the skewness:

$$Sk_\mathrm{a} = \frac{\left\langle (a - \langle a \rangle)^3 \right\rangle}{\sigma^3} \,, \tag{2.33}$$

and the kurtosis:

$$Ku_\mathrm{a} = \frac{\left\langle (a - \langle a \rangle)^4 \right\rangle}{\sigma^4} \,. \tag{2.34}$$

Remember that for a variable characterized by the Gaussian distribution, $Sk_\mathrm{a} = 0$ and $Ku_\mathrm{a} = 3$, being $a_1 = -\infty$ and $a_2 = \infty$.

Similar definitions for the moments can be given for a pair of stochastic variables α and β with probability density function $f_{\alpha\beta}(a, b)$. In particular, the covariance reads:

$$\int_{a_1}^{a_2} \int_{b_1}^{b_2} (a - \langle a \rangle)(b - \langle b \rangle) f_{\alpha\beta}\, \mathrm{d}a\, \mathrm{d}b = \langle ab \rangle - \langle a \rangle \langle b \rangle \,. \tag{2.35}$$

In many practical cases we have a discrete approximation of the probability density function, i.e. an histogram, and the stochastic variable is discretized. Thus the integrals become sums, but the essentials aspects remain the same.

2.3.2 *Averaging*

Any stochastic variable $\tilde{a}$ (for instance, the velocity components, or the temperature) can be represented as the sum of the mean value plus a fluctuation:

$$\tilde{a} = \langle \tilde{a} \rangle + a = A + a \,, \tag{2.36}$$

and $\langle a \rangle = 0$. See Finnigan and Shaw (2008) for a detailed discussion about averaging.

The statistical moments are often computed in a fixed reference frame (space coordinates and time) and describe the dynamic and state variable fields: the Eulerian description of the flow. Moments can also be computed following the fluid parcel motion: the Lagrangian description.

2.3.2.1 Ensemble Average

The ensemble average implies the availability of a number of different realizations of the same process, so that the statistical properties of this ensemble of realizations are investigated (Monin and Yaglom 1971). Ensemble averaging is the formally correct way to derive equations for the moments from (2.9), (2.10), (2.14) and (2.27) (or any other equation discussed above).

Ensemble averaging is in principle possible with laboratory observations, only in an approximate way with geophysical observations. With some caveats, it can be made using the numerical solutions of the governing equations.

The ensemble average of N realizations of the variable $\tilde{a}$ is:

$$\langle \tilde{a} \rangle (\boldsymbol{x}, t) = \frac{1}{N} \sum_{i=1}^{N} \tilde{a}^{(i)}(\boldsymbol{x}, t) \ , \tag{2.37}$$

where $\tilde{a}^{(i)}$ is the ith realization. This definition is the discrete equivalent to (2.31) with $n = 1$, including the normalization condition.

The following rules hold:

$$\langle \tilde{a} + \tilde{b} \rangle = \langle \tilde{a} \rangle + \langle \tilde{b} \rangle \ , \tag{2.38}$$

and

$$\left\langle \frac{\partial \tilde{a}}{\partial x} \right\rangle = \frac{\partial \langle \tilde{a} \rangle}{\partial x} \ , \tag{2.39}$$

for any independent variable x (derivation commutes with averaging).

2.3.2.2 Time Average

Field observations are in general made in a fixed point (the position of the instrument) and the independent variable is time. The time average is

$$\langle \tilde{a} \rangle (\boldsymbol{x}, \vartheta) = \frac{1}{T_{\mathrm{A}}} \int_{\vartheta - T_{\mathrm{A}}/2}^{\vartheta + T_{\mathrm{A}}/2} \tilde{a}(\boldsymbol{x}, t) \, \mathrm{d}t \ , \tag{2.40}$$

being T_{A} the averaging time. If the variable $\tilde{a}(\boldsymbol{x}, t)$ is statistically steady, the averaging time can be chosen long enough to average over a representative sample of realizations (the time average converges to ensemble average at increasing averaging time). Field observations are in general unsteady: thus T_{A} must be chosen smaller than the characteristic time of occurrence of unsteadiness, and long enough to ensure the statistical representativeness of the sample.

The commutative property holds with respect to the time derivative (see Exercise 2.4).

Note that (2.40) is a filter which operates on the original signal. The time averaging means that we neglect the high frequency variability of the signal, which is represented by the moments of the fluctuations about the mean: it must be always kept in mind the relevance of the length of the averaging time in the investigation of turbulent phenomena. Note also that in practice the averaged variable $\langle \tilde{a} \rangle$ is discretized in time at intervals T_A (although from the definition it is continuous). A further discussion can be found in the Exercise 2.5.

2.3.2.3 Volume Average

In numerical modelling, the discretization in space implies that any variable is averaged over a finite volume, and this implies a filtering in the wavenumber spectrum (which will be discussed later). The volume average was introduced in connection with the early numerical weather simulations (Deardorff 1970); see also Pope (2000, p. 576).

The volume averaged variable $\langle \tilde{a} \rangle(\chi_1, \chi_2, \chi_3, t)$ is the mean value in a box (parallelepiped) of sides L_{A1}, L_{A2}, L_{A3}. Thus

$$\langle \tilde{a} \rangle(\chi_1, \chi_2, \chi_3, t) = \frac{1}{L_{A1} L_{A2} L_{A3}} \int_{\chi_1 - L_{A1}/2}^{\chi_1 + L_{A1}/2} \int_{\chi_2 - L_{A2}/2}^{\chi_2 + L_{A2}/2} \int_{\chi_3 - L_{A3}/2}^{\chi_3 + L_{A3}/2} \tilde{a}(\boldsymbol{x}, t)\, \mathrm{d}x_1\, \mathrm{d}x_2\, \mathrm{d}x_3 \qquad (2.41)$$

The commutative property holds as for time averaging. An interesting issue arises if the volume contains obstacles (like in a canopy): in this case the commutative property is not satisfied for all the variables, as will be discussed for the urban boundary layers and for flows into the vegetative canopies (Sect. 4.6).

2.3.3 *Covariances and Spectra*

Let consider two variables $\tilde{a} = A + a$ and $\tilde{b} = B + b$, both functions of the position $\boldsymbol{x}$ and of the time t. The covariance of the fluctuations is given by

$$R_{ab}(\boldsymbol{x}, \boldsymbol{r}, t, \tau) = \langle a(\boldsymbol{x}, t)\, b(\boldsymbol{x} + \boldsymbol{r}, t + \tau) \rangle \ , \qquad (2.42)$$

where $\boldsymbol{r}$ is the separation and τ the time lag.

2.3.3.1 Eulerian Covariances

In an Eulerian framework, we will consider space covariances at fixed time (i.e. $\tau = 0$), or time covariances in a fixed position (i.e. $\boldsymbol{r} = 0$). Moreover, we shall refer to homogeneous conditions (i.e. the statistical properties are independent on position) for space covariances, so that they depend on $r = |\boldsymbol{r}|$ only, or to steady conditions for time covariances, so they depend on $|\tau|$ only.

It is straightforward to obtain Eulerian time covariances from data recorded in a fixed point (although the requirement of steadiness implies some care in the choice of the time intervals). Space covariances between velocity components can be obtained in laboratory, using tracers and image analysis and in the field using arrays of instruments or along the wind direction under the 'frozen turbulence' approximation, described in Sect. 2.3.3.4 below.

2.3.3.2 Lagrangian Covariances

Also Lagrangian covariances will be used, in order to correlate the dynamic variables along the trajectories of the fluid parcels.

Let refer to homogeneous and steady conditions. If we consider a single parcel, and two positions along its trajectory, the time covariance reduces to a function of the elapsed time (from a fixed origin) and of $|\tau|$. For two parcels, whose trajectories are described by $\boldsymbol{x}(t)$ and by $\boldsymbol{x}(t) + \boldsymbol{r}(t)$, the time covariance $R_{\text{ab}}[\tau, r(t, \tau)]$ depends on the elapsed time, on the absolute value of the time lag and on the separation $r = |\boldsymbol{x}(t + \tau) + \boldsymbol{r}(t + \tau) - \boldsymbol{x}(t)|$.

2.3.3.3 Integral Scales

We require that in steady and homogeneous conditions the integral of the covariance between 0 and ∞ converges. Thus, at fixed time:

$$L_{\text{ab}}^{(k)} = \frac{1}{\langle ab \rangle} \int_0^\infty R_{\text{ab}}(r_k)\, \mathrm{d}r_k \tag{2.43}$$

defines the integral length scale in the k direction. Similarly:

$$T_{\text{ab}} = \frac{1}{\langle ab \rangle} \int_0^\infty R_{\text{ab}}(\tau)\, \mathrm{d}\tau \tag{2.44}$$

defines the integral time scale.

It is important to note that using the observations, the convergence of the numerical discretization of the integral is not ensured, because phenomena with different space or time scales are usually superimposed.

2.3.3.4 The 'Frozen Turbulence' Approximation

The 'frozen turbulence' approximation allows to use the time correlation to derive the space correlation in the mean wind direction. The idea is discussed by Wyngaard and Coté (1971, p. 191) referring to Heskestad (1965). See also the recent discussion by Higgins et al. (2012). Let write the 'frozen turbulence' hypothesis as

$$\frac{\mathrm{D}a}{\mathrm{D}t} = 0\,, \tag{2.45}$$

i.e. we make the hypothesis that the variable a in the turbulent flow is advected without modifications: this approximation holds in general if the time and space intervals taken under consideration are small enough with respect to the typical scales of changes.

With a proper choice of the axes the mean wind is $(U, 0, 0)$. It results:

$$\frac{\partial a}{\partial t} = -\left[(U+u)\frac{\partial a}{\partial x} + v\frac{\partial a}{\partial y} + w\frac{\partial a}{\partial z}\right]. \tag{2.46}$$

Assuming that U is steady, that the scale of the velocity fluctuations is much smaller than the mean velocity: $\langle u_i^2 \rangle \ll U^2$ for any i, and retaining only the first-order terms, (2.46) becomes:

$$\frac{\partial a}{\partial x} = -\frac{1}{U}\frac{\partial a}{\partial t}\,, \tag{2.47}$$

and can be integrated giving $a(x_2) - a(x_1) = a(t_1) - a(t_2)$, being $x = Ut$, that is, the space covariance along the mean wind direction can be computed from time measurements in a fixed point.

This approximation holds if the turbulent eddies are transported by the mean flow over the measuring point with minor distortions, so two observations taken with a τ lag correspond to two observations taken with separation $r = U\,\tau$ along the mean wind direction. It results, for time lags between 0 and τ_{max}:

$$R_{aa}(r) = R_{aa}(\tau) = \langle a(0)a(\tau)\rangle = \frac{1}{T_A}\int_0^{T_A} a(t)a(t+\tau)\,\mathrm{d}t\,,\ \tau = r/U\,, \tag{2.48}$$

where $T_A \gg T_{aa}$ and the total record length must be $T_A + \tau_{max}$.

2.3.3.5 Averaging Time and Correlation Time

The finiteness of the averaging time is unavoidable, and the error in the estimate of the average is related to the integral time scale (see Wyngaard 2010, p. 35). Let consider a statistically steady variable $\tilde{a}(t) = A + a(t)$, where A is the true average value, say the average derived from a very long (infinite) averaging time, and its time average $A^{(T_A)}(\vartheta)$ obtained from a finite averaging time T_A, according to (2.40). The difference between the two averages reads:

$$A^{(T_A)}(\vartheta) - A = \frac{1}{T_A} \int_{\vartheta - T_A/2}^{\vartheta + T_A/2} a(t)\, dt\,, \tag{2.49}$$

i.e., the difference between the finite time average and the true average is related to the average of the fluctuations in the averaging time interval (this average is in general different from zero).

Because of the steadiness of the process, the variance of this difference, σ^2, does not depend on ϑ. It reads:

$$\begin{aligned} \sigma^2 &\equiv \left\langle \left(A^{(T_A)}(\vartheta) - A\right)^2 \right\rangle \\ &= \frac{1}{T_A^2} \int_{\vartheta - T_A/2}^{\vartheta + T_A/2} dt_1 \int_{\vartheta - T_A/2}^{\vartheta + T_A/2} dt_2 \langle a(t_1) a(t_2) \rangle\,. \end{aligned} \tag{2.50}$$

According to the definition (2.42), the term in brackets is the time covariance of the fluctuations $\langle a(t_1)a(t_2)\rangle = R_{aa}(|t_1 - t_2|)$, which is a function of the absolute value of the time difference. The double integral is transformed in a single one (see Exercise 2.6) and the variance reads:

$$\sigma^2 = \frac{2}{T_A^2} \int_{\vartheta - T_A/2}^{\vartheta + T_A/2} R_{aa}(t)(T_A - t)\, dt\,. \tag{2.51}$$

Thus the difference between the true average and the average obtained from a finite averaging time depends on the values of the averaging time and of the integral time scale T_{aa} of the fluctuations, as from (2.44). For averaging times larger than the correlation time, it results:

$$\sigma^2 \approx 2\langle a^2 \rangle \frac{T_{aa}}{T_A}\,, \tag{2.52}$$

i.e., the variance of the difference between the finite time average and the true average decreases as the ratio between the integral time scale and the averaging time decreases. In unsteady cases (like any field experiment), reliable estimates of the mean can be obtained if the time characterizing the unsteadiness is much larger than the correlation time.

2.3.3.6 Cospectra, Power Spectra

We use the following definitions for the Fourier transforms:

$$\mathcal{F}[f(x);\chi] \equiv \hat{f}(\chi) = \frac{1}{2\pi}\int_{-\infty}^{+\infty} f(x)\exp(-i\chi x)\,\mathrm{d}x\,, \tag{2.53}$$

and

$$f(x) = \int_{-\infty}^{+\infty} \hat{f}(\chi)\exp(+ix\chi)\,\mathrm{d}\chi\,, \tag{2.54}$$

where x is the separation or the time lag and χ is the wavenumber or the frequency, respectively. For real, even functions, (2.53) and (2.54) reduce to cosine transforms.

Define the cospectrum, or power spectrum, as twice the cosine transform of the covariance:

$$E_{\mathrm{ab}}(\chi) = \frac{2}{\pi}\int_{0}^{\infty} R_{\mathrm{ab}}(x)\cos(x\chi)\,\mathrm{d}x\,. \tag{2.55}$$

Thus the covariance results:

$$R_{\mathrm{ab}}(x) = \int_{0}^{\infty} E_{\mathrm{ab}}(\chi)\cos(x\chi)\,\mathrm{d}\chi\,. \tag{2.56}$$

Note that, from (2.56):

$$R_{\mathrm{ab}}(0) \equiv \langle a\,b\rangle = \int_{0}^{\infty} E_{\mathrm{ab}}(\chi)\,\mathrm{d}\chi\,, \tag{2.57}$$

i.e., $E_{\mathrm{ab}}(\chi)\,\mathrm{d}\chi$ can be interpreted as the contribution from the wavenumber or frequency range $[\chi, \chi + \mathrm{d}\chi]$ to the total covariance.

2.3.3.7 Average and Spectra

Averaging procedure corresponds to filter the spectrum (see Pope 2000, p. 569). To understand the effect of the time averaging, we rewrite (2.40) as follows:

$$\langle\tilde{a}\rangle(\vartheta) \equiv \frac{1}{T_{\mathrm{A}}}\int_{\vartheta-T_{\mathrm{A}}/2}^{\vartheta+T_{\mathrm{A}}/2} \tilde{a}(t)\,\mathrm{d}t = \int_{-\infty}^{\infty} G(t)\tilde{a}(\vartheta - t)\,\mathrm{d}t\,, \tag{2.58}$$

where G is the top-hat filter

$$G = \frac{1}{T_{\mathrm{A}}} \quad \text{for } |t| \le T_{\mathrm{A}}/2 ,$$
$$G = 0 \quad \text{for } |t| > T_{\mathrm{A}}/2 . \tag{2.59}$$

Note in passing that (2.58) and (2.59) are equivalent to write

$$\langle \tilde{a} \rangle (\vartheta) = \int_{-\infty}^{\infty} G(\vartheta - t)\tilde{a}(t)\,\mathrm{d}t , \tag{2.60}$$

with the obvious change $G = 1/T_{\mathrm{A}}$ for $|\vartheta - t| \le T_{\mathrm{A}}/2$.

It results (see Exercise 2.5) that the power spectrum $E_{\mathrm{aa}}^{(F)}(\omega)$ of the averaged (filtered) quantity is:

$$E_{\mathrm{aa}}^{(F)}(\omega) = \left[\frac{\sin(\omega T_{\mathrm{A}}/2)}{\omega T_{\mathrm{A}}/2}\right]^2 E_{\mathrm{aa}}(\omega) , \tag{2.61}$$

i.e., there is a reduction of the amplitude of the high-frequency part of the spectrum (although the top-hat filter doesn't produce a sharp cut-off of the spectrum). Figure 2.1 shows an evident reduction for $\omega T_{\mathrm{A}} \sim 6$: the cut-off period is of the order of the averaging time.

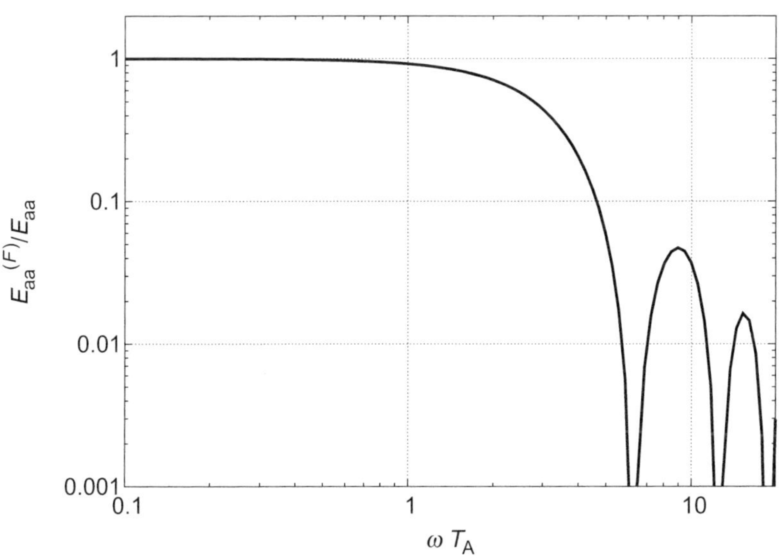

Fig. 2.1 Ratio between the filtered and the unfiltered power spectra, as function of the nondimensional frequency ωT_{A}, from (2.61)

2.3.3.8 A Remark About Notation

When dealing with the velocity, we shall use R_{ij} for the covariance between the components u_i and u_j, and E_{ij} for the corresponding spectrum. Similarly, we shall use L_{ij} and T_{ij} for the integral length and time scales. The subscript uu will refer to a generic component.

2.4 Reynolds Averaged Equations

To obtain the equations for the statistical moments of the variables, we will refer to the ensemble average. Volume average will be considered for numerical applications, and will be discussed later.

Hereinafter incompressible flows will be considered, if not explicitly stated otherwise.

2.4.1 The Equations for the First-Order Moments

According to (2.36), we shall split the variables in the average plus the fluctuation: $\tilde{u}_i = U_i + u_i$, $\tilde{\theta} = \Theta + \theta$ and $\tilde{c} = C + c$. Taking the ensemble average of (2.9), where, according to the last sentence in Sect. 2.2.6, the potential temperature has been substituted to the temperature, the equation for the mean velocity components reads:

$$\frac{\mathrm{D}U_i}{\mathrm{D}t} = -\frac{1}{\rho_{00}}\frac{\partial\langle\tilde{p}\rangle}{\partial x_i} + \varepsilon_{ij3} f U_j + \nu\frac{\partial^2 U_i}{\partial x_j \partial x_j} + \delta_{i3}\frac{g}{\theta_{00}}\Theta - \frac{\partial\langle u_i u_j\rangle}{\partial x_j}\,, \tag{2.62}$$

where the material derivative is written in terms of the mean velocity U_j:

$$\frac{\mathrm{D}}{\mathrm{D}t} \equiv \frac{\partial}{\partial t} + U_j\frac{\partial}{\partial x_j}\,. \tag{2.63}$$

The continuity equation (2.11) reads:

$$\frac{\partial U_i}{\partial x_i} = 0\,. \tag{2.64}$$

The temperature equation (2.27) becomes:

$$\frac{\mathrm{D}\Theta}{\mathrm{D}t} + U_3\frac{\mathrm{d}\theta_0}{\mathrm{d}x_3} = \kappa_{\mathrm{T}}\frac{\partial^2\Theta}{\partial x_j \partial x_j} - \frac{\partial\langle u_j\theta\rangle}{\partial x_j}\,, \tag{2.65}$$

where θ_0 is assumed to be function of z only.

The equation for the concentration of a passive scalar (2.15) reads:

$$\frac{\mathrm{D}C}{\mathrm{D}t} = \kappa_{\mathrm{c}} \frac{\partial^2 C}{\partial x_j \partial x_j} - \frac{\partial \langle u_j c \rangle}{\partial x_j} . \tag{2.66}$$

As discussed before, for high Reynolds number, the viscous term can be neglected in (2.62). In order to neglect the analogous terms in (2.65) and (2.66), it is necessary that the Prandtl number $P_{\mathrm{r}} = \nu / \kappa_{\mathrm{T}}$ and the Schmidt number $S_{\mathrm{c}} = \nu / \kappa_{\mathrm{c}}$ are finite of order one.

Note that these equations are not closed, because of the presence of moments of the second order, i.e. the covariances between fluctuations, also called turbulent fluxes. As well known, this problem exists for the equations of the moments of any degree n (which contain moments of order $n+1$), and the solution requires a closure, that is an empirical hypothesis which expresses the nth-order moments in terms of the $(n-1)$th-order ones. We will face this problem throughout the study, and it will be discussed in detail in the last part of the book.

2.4.1.1 The Geostrophic Wind

Let us write the mean pressure as a large scale contribution, whose horizontal gradients are constant, and a small scale contribution, due to the dynamics of the PBL itself: $\langle \tilde{p} \rangle = P_{\mathrm{g}} + P$. Note that when we will deal with homogeneous conditions (at the scale of PBL), the horizontal gradient of P goes to zero, while the gradient of P_{g} remains: it is a forcing term.

In absence of PBL (small scale) dynamics, (2.62) allows to define the geostrophic wind: $\boldsymbol{U}_{\mathrm{g}} = (U_{\mathrm{g}}, V_{\mathrm{g}}, 0)$:

$$\begin{aligned} -\frac{1}{\rho_{00}} \frac{\partial P_{\mathrm{g}}}{\partial x} + f V_{\mathrm{g}} &= 0 , \\ -\frac{1}{\rho_{00}} \frac{\partial P_{\mathrm{g}}}{\partial y} - f U_{\mathrm{g}} &= 0 . \end{aligned} \tag{2.67}$$

Here the geostrophic wind is considered as a basic state of motion, produced by the large scale pressure gradients. Note that a formal introduction of the geostrophic wind can be found, for instance, in Gill (1982) and in general in any textbook of meteorology.

Thus, as far the horizontal components of the velocity are concerned, (2.62) reads:

$$\frac{\mathrm{D}U_i}{\mathrm{D}t} = -\frac{1}{\rho_{00}} \frac{\partial P}{\partial x_i} + \varepsilon_{ij3} f \left(U_j - U_{\mathrm{g}j} \right) + \nu \frac{\partial^2 U_i}{\partial x_j \partial x_j} - \frac{\partial \langle u_i u_j \rangle}{\partial x_j}, \; i = 1, 2 . \tag{2.68}$$

2.4.2 The Equations for the Fluctuations

The equation for the fluctuations of a given quantity is obtained by subtracting the equation for the mean quantity from the corresponding equation for the total quantity: for instance, subtracting (2.62) from (2.9) we obtain the equation for u_i.

In this study we shall use only the continuity equation for the velocity fluctuations in a constant density flow:

$$\frac{\partial u_i}{\partial x_i} = 0 \,, \tag{2.69}$$

i.e., the divergence of the velocity fluctuations is zero.

2.4.3 The Equations for the Second-Order Moments of Velocity

By multiplying the equation for the fluctuations u_i by u_k, summing it by the equation for u_k multiplied by u_i and taking the average, the equation for the second-order moments of the velocity components is derived:

$$\begin{aligned}\frac{\mathrm{D}\langle u_i u_k\rangle}{\mathrm{D}t} &= \left(-\langle u_i u_j\rangle\frac{\partial\langle u_k\rangle}{\partial x_j} - \langle u_k u_j\rangle\frac{\partial\langle u_i\rangle}{\partial x_j}\right) - \frac{\partial\langle u_i u_j u_k\rangle}{\partial x_j}\\ &+ \frac{g}{\theta_{00}}(\delta_{k3}\langle u_i\theta\rangle + \delta_{i3}\langle u_k\theta\rangle) + f(\varepsilon_{kj3}\langle u_i u_j\rangle + \varepsilon_{ij3}\langle u_k u_j\rangle)\\ &- \frac{1}{\rho_{00}}\left\langle u_k\frac{\partial p}{\partial x_i} + u_i\frac{\partial p}{\partial x_k}\right\rangle\\ &+ \nu\left\langle u_k\frac{\partial^2 u_i}{\partial x_j\partial x_j} + u_i\frac{\partial^2 u_k}{\partial x_j\partial x_j}\right\rangle \,.\end{aligned} \tag{2.70}$$

The meaning of the various terms at the RHS of the previous equation is:

- 1: production by effect of the coupling between covariances and mean shear;
- 2: divergence of third order moments, i.e., transport of covariances
- 3: production/destruction by buoyancy
- 4: Coriolis effect
- 5: transfer due to the coupling between fluctuations of pressure and velocity.This term may be rewritten as follows:

$$-\frac{1}{\rho_{00}}\left(\frac{\partial\langle p u_k\rangle}{\partial x_i} + \frac{\partial\langle p u_i\rangle}{\partial x_k}\right) - \left\langle\frac{p}{\rho_{00}}\left(\frac{\partial u_i}{\partial x_k} + \frac{\partial u_k}{\partial x_i}\right)\right\rangle \,. \tag{2.71}$$

Note that this term is of order three, like term 2;

- 6: molecular dissipation. This term may be rewritten as follows:

$$\nu \frac{\partial^2 \langle u_i u_k \rangle}{\partial x_j \partial x_j} - 2\nu \left\langle \frac{\partial u_i}{\partial x_j} \frac{\partial u_k}{\partial x_j} \right\rangle . \tag{2.72}$$

The first term in (2.72) represents the molecular diffusion of the second-order moment $\langle u_i u_k \rangle$, while the second one is proportional to the covariance of the derivatives of the velocity fluctuations and represents the turbulent dissipation due to viscosity. An evaluation of its magnitude will be made in Sect. 2.5 for a specific paradigm.

2.4.3.1 The Equation for the Mean Turbulent Kinetic Energy (TKE)

Putting $i = k$ in (2.70), the equation for the mean turbulent kinetic energy $E_\mathrm{k} = \frac{1}{2}\langle u_i u_i \rangle$ results:

$$\frac{\mathrm{D}E_\mathrm{k}}{\mathrm{D}t} = -\langle u_i u_j \rangle \frac{\partial U_i}{\partial x_j} - \frac{\partial}{\partial x_j}\left(\frac{\langle p u_j \rangle}{\rho_{00}} + \frac{1}{2}\langle u_i u_i u_j \rangle \right)$$
$$+ \frac{g}{\theta_{00}} \langle \theta u_3 \rangle + \nu \frac{\partial^2 E_\mathrm{k}}{\partial x_j \partial x_j} - \varepsilon . \tag{2.73}$$

Note that

$$\varepsilon = \nu \left\langle \frac{\partial u_i}{\partial x_j} \frac{\partial u_i}{\partial x_j} \right\rangle \tag{2.74}$$

is a positive term which describes the viscous dissipation of the TKE.

2.4.4 *The Equation for the Temperature Variance*

The equation for the temperature variance is obtained by multiplying the equation for the temperature fluctuations by the temperature fluctuations themselves and taking the average. It results:

$$\frac{\mathrm{D}\langle \theta^2 \rangle}{\mathrm{D}t} = -2\langle u_i \theta \rangle \frac{\partial \Theta_\mathrm{f}}{\partial x_i} - \frac{\partial \langle u_i \theta^2 \rangle}{\partial x_i} + \kappa_\mathrm{H} \left\langle \theta \frac{\partial^2 \theta}{\partial x_i \partial x_i} \right\rangle , \tag{2.75}$$

where $\Theta_\mathrm{f} = \theta_{00} + \theta_0 + \Theta$ is the average temperature of the flow. The last term may be rewritten, in a similar way to (2.72), as the sum of the viscous diffusion of the fluctuation variance and the viscous dissipation:

$$\kappa_{\mathrm{H}}\left\langle\theta\frac{\partial^2\theta}{\partial x_i\partial x_i}\right\rangle = \kappa_{\mathrm{H}}\frac{\partial^2\langle\theta^2\rangle}{\partial x_i\partial x_i} - 2\varepsilon_\theta\ , \tag{2.76}$$

where

$$\varepsilon_\theta = \kappa_{\mathrm{H}}\left\langle\left(\frac{\partial\theta}{\partial x_j}\right)^2\right\rangle \tag{2.77}$$

is the (positive) dissipation term.

Note that the term $-\langle u_3\theta\rangle\partial\Theta_{\mathrm{f}}/\partial x_3 \geq 0$ always represents the production of temperature variance (the vertical production term).

For any passive tracer, the equation for the fluctuation variance is the same as (2.75), by substituting θ with c and κ_{H} with κ_{c}. The dissipation term is analogous to (2.77).

2.4.5 *The Equations for the Heat Fluxes*

These equations are derived in a similar way as those for the momentum fluxes. They read:

$$\begin{aligned}\frac{\mathrm{D}\langle u_i\theta\rangle}{\mathrm{D}t} &= -\langle u_iu_j\rangle\frac{\partial\Theta}{\partial x_j} - \langle u_j\theta\rangle\frac{\partial U_i}{\partial x_j} - \frac{\partial\langle u_iu_j\theta\rangle}{\partial x_j} - \frac{1}{\rho_{00}}\left\langle\theta\frac{\partial p}{\partial x_i}\right\rangle + \delta_{i3}\frac{g}{\theta_{00}}\langle\theta^2\rangle \\ &+ \nu\left\langle\theta\frac{\partial^2 u_i}{\partial x_jx_j}\right\rangle + \kappa_{\mathrm{H}}\left\langle u_i\frac{\partial^2\theta}{\partial x_jx_j}\right\rangle\ .\end{aligned} \tag{2.78}$$

The two last terms are proportional to the viscosity and to the thermal diffusivity. Each can be rewritten, as done in (2.72), as the sum of a Laplacian of the flux and the covariance of the derivatives, giving the following expression:

$$\frac{\nu+\kappa_{\mathrm{H}}}{2}\frac{\partial^2\langle u_i\theta\rangle}{\partial x_j\partial x_j} - (\nu+\kappa_{\mathrm{H}})\left\langle\frac{\partial u_i}{\partial x_j}\frac{\partial\theta}{\partial x_j}\right\rangle\ , \tag{2.79}$$

where the first term (molecular diffusion of the turbulent flux) can be neglected for high Reynolds number flows, while the second one represents the turbulent dissipation at small scales.

2.4.6 The Interpretation of the Fluctuation Covariances and the Eddy Diffusion Model

The covariances between the velocity components can be interpreted as turbulent momentum fluxes, and those between temperature and velocity as heat fluxes. In this section, fluxes are related by an heuristic argument to the mean gradients, leading to an expression for the eddy diffusion coefficients in the simple configuration illustrated in Fig. 2.2; more general formulations, like those discussed in the Chap. 6, are direct extensions of this one.

Let consider a velocity field $[U(z) + u, v, w]$, with $\mathrm{d}U/\,\mathrm{d}z > 0$. For a small vertical displacement l, if $w > 0$, the horizontal fluctuation turns out to be negative: $u < 0$; the opposite occurs if $w < 0$. On average, $\langle uw \rangle < 0$. According to the flux-gradient paradigm, $\langle uw \rangle$ is interpreted as the kinematic vertical momentum flux, proportional to minus the mean velocity gradient, through a eddy diffusion coefficient K_τ for momentum:

$$\langle uw \rangle = -K_\tau \frac{\mathrm{d}U}{\mathrm{d}z} \,. \tag{2.80}$$

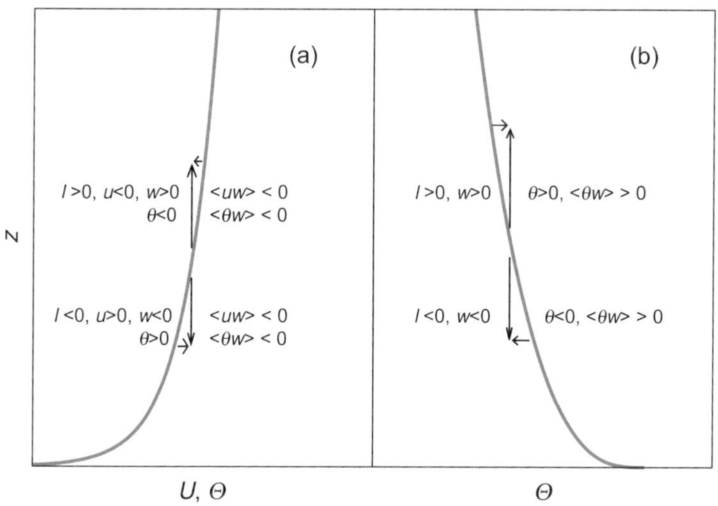

Fig. 2.2 Interpretation of the covariances. In (**a**) the case of the mean quantity (velocity or temperature) increasing with height is represented: the covariance between u and w, or between θ and w (the turbulent vertical flux of momentum or heat, respectively) is negative (flux downwards). In (**b**) the case of mean temperature decreasing with height is represented: the covariance is positive (vertical flux of heat upwards)

To determine K_τ, a simple approach starts from:

$$u = -[U(z+l) - U(z)] \approx -l\frac{\mathrm{d}U}{\mathrm{d}z}\,, \tag{2.81}$$

so that

$$\langle uw \rangle = -\langle lw \rangle \frac{\mathrm{d}U}{\mathrm{d}z}\,, \tag{2.82}$$

which gives an expression for the eddy diffusion coefficient. Putting:

$$w \approx -\alpha \mathrm{sgn}\left(\frac{\mathrm{d}U}{\mathrm{d}z}\right) u = \alpha l \mathrm{sgn}\left(\frac{\mathrm{d}U}{\mathrm{d}z}\right)\frac{\mathrm{d}U}{\mathrm{d}z}\,, \tag{2.83}$$

and substituting in (2.82), it results:

$$K_\tau = \langle \alpha l^2 \rangle \left|\frac{\mathrm{d}U}{\mathrm{d}z}\right|\,. \tag{2.84}$$

The square root of $\langle \alpha l^2 \rangle$ is called the mixing length l_τ and it is expected to be of the order of the correlation scale $L_{\mathrm{uw}}^{(z)}$ as defined in (2.43).

In presence of vertical gradients of both velocity and potential temperature, a similar approach leads to recognize that

$$\langle \theta w \rangle = -K_{\mathrm{Q}}\frac{\mathrm{d}\Theta}{\mathrm{d}z}\,, \tag{2.85}$$

where in this simple scheme the eddy diffusion coefficient of heat is equal to that of momentum. Analogous expressions can be derived for any scalar tracer.

It must be observed that if w is correlated to u and to θ, also u and θ are correlated: see Monin and Yaglom (1971, pp. 523, 613, 666) or Wyngaard et al. (1971, Fig. 4), and thus even in absence of any horizontal gradient of mean temperature, a horizontal heat flux arises, and its sign depends on the sign of the vertical gradient of temperature. The heuristic argument outlined above gives:

$$\langle \theta u \rangle = \langle l^2 \rangle \frac{\mathrm{d}U}{\mathrm{d}z}\frac{\mathrm{d}\Theta}{\mathrm{d}z}\,, \tag{2.86}$$

which suggests that the horizontal heat flux is related to the vertical heat flux. This point will be made more quantitative in Sect. 3.4.3.

The existence of a flux in absence of a mean gradient in the same direction highlights the limits of validity of the flux-gradient relationship in the turbulent flows. On the other hand, this relationship represents a powerful tool to close the equations, and can be used safely in many different situations. Because the eddy diffusion coefficients (eddy viscosity) are features of the flow, and thus depend on the dynamics, they must be prescribed on the basis of the dynamics itself: to close the equations, it is necessary to guess how the solution of the equations is. A practical way to do this in a time integration, is to use the informations of the previous time step.

2.5 Universal Features of Shear-Dominated Turbulence

The inertial subrange theory (Kolmogorov 1941) gives the interpretation paradigm for shear-dominated turbulence, in short K41.

From the NSE (2.9), under the conditions of negligible external forces, of negligible viscosity (i.e., in the limit $R_e \rightarrow \infty$), and far from the boundaries, it results:

$$\frac{\mathrm{D}\tilde{u}_i}{\mathrm{D}t} = -\frac{1}{\rho_{00}}\frac{\partial \tilde{p}}{\partial x_i}, \tag{2.87}$$

which is invariant under arbitrary changes to the length and velocity scales $x \rightarrow \lambda x$ and $\tilde{u}_i \rightarrow \lambda^h \tilde{u}_i$ corresponding to $t \rightarrow \lambda^{1-h} t$. Remember that $\tilde{p}/\rho_{00}$ scales as $\tilde{u}^2$.

This scale invariance means that the dynamics described by (2.87) is the same in a range of scales, which must be smaller than the space scale of the domain (the size of the container of the fluid) and large enough to ensure that the Reynolds number is greater than unity (i.e. scales larger than the scale at which viscosity becomes important). This range is called inertial subrange. For our purposes, the central hypothesis is that the inertial subrange is characterized by isotropy and scale-independent dissipation of the TKE.

From the TKE equation (2.73), under the same conditions, it results:

$$\frac{\mathrm{D}E_\mathrm{k}}{\mathrm{D}t} = -\langle u_i u_j \rangle \frac{\partial U_i}{\partial x_j} - \frac{\partial}{\partial x_j}\left(\frac{\langle p u_j \rangle}{\rho_{00}} + \frac{1}{2}\langle u_i u_i u_j \rangle \right) - \lambda^{3h-1}\varepsilon\,, \tag{2.88}$$

which leads to $h = 1/3$ in order to guarantee that the dissipation is scale-independent.

The dimensional analysis and the scale-independence requirement indicate that

$$\varepsilon \propto \frac{\mathrm{U}_\mathrm{l}^3}{\mathrm{l}}, \tag{2.89}$$

for any value of the length l in the inertial subrange, being U_l the velocity scale at the same length. Thus:

$$\mathrm{U}_\mathrm{l}^2 \propto \varepsilon^{2/3}\mathrm{l}^{2/3}\,. \tag{2.90}$$

This relation allows us to estimate the scale of the velocity derivatives and thus of the dissipation term (2.74) in the TKE equation. In fact, $|\partial u/\partial x| \approx \mathrm{U_l}/\mathrm{l} \propto \mathrm{l}^{-2/3}$, that is, the smaller scales are expected to contribute more than the larger ones to the dissipation. For this reason the dissipation, although related to the molecular viscosity, cannot be neglected even in high Reynolds number flows.

The lower boundary of the inertial subrange $\mathrm{l_K}$ may be estimated assuming that the Reynolds number at that scale is $O(1)$:

$$1 \sim \frac{\mathrm{l_K U_{l_K}}}{\nu} = \frac{\varepsilon^{1/3}\mathrm{l_K^{4/3}}}{\nu} \,. \tag{2.91}$$

Thus $\mathrm{l_K}$, called Kolmogorov length scale, results:

$$\mathrm{l_K} = \nu^{3/4}\varepsilon^{-1/4} \,, \tag{2.92}$$

and the related time scale reads:

$$\mathrm{T_K} = (\nu/\varepsilon)^{1/2} \,. \tag{2.93}$$

Assuming that the upper boundary of the inertial subrange is approximated by the integral length scale L of a velocity component (2.43), where we shall neglect subscript and superscript because of the isotropy of the flow, a velocity scale U_L can be evaluated from (2.90), and the width of the inertial subrange results to be a function of the Reynolds number based on the integral scales $R_\mathrm{e} = \mathrm{U_L}L/\nu$:

$$\frac{L}{\mathrm{l_K}} = \frac{L}{\varepsilon^{-1/4}\nu^{3/4}} = \frac{L\mathrm{U_L^{3/4}}L^{-1/4}}{\nu^{3/4}} = R_\mathrm{e}^{3/4} \,, \tag{2.94}$$

which gives an estimate of the width of the inertial subrange in terms of length scales. In practice, this means that to describe the entire inertial subrange in a numerical simulation, the number of grid points in each direction increases as the Reynolds number increases, according to (2.94).

2.5.1 *Velocity Covariances and Spectra*

Let simplify the analysis putting the mean velocity U constant, for instance equal to zero.

2.5.1.1 Eulerian Description

The second-order structure functions reads:

$$D_{\mathrm{uu}}(r) = \left\langle [u(r) - u(0)]^2 \right\rangle = 2\left(\left\langle u^2 \right\rangle - R_{\mathrm{uu}}(r)\right) , \tag{2.95}$$

where R_{uu} is the Eulerian covariance as from (2.42) and the subscript uu refers to a generic component of the velocity. Hereinafter r will be taken as the separation in the direction of u. In the inertial subrange it must be:

$$D_{\mathrm{uu}}(r) \approx C_{\mathrm{K}} \varepsilon^{2/3} r^{2/3} , \tag{2.96}$$

where $C_{\mathrm{K}} = 2$ (see Ouellette et al. 2006).

Remember in passing the following exact relationship:

$$D_{\mathrm{uuu}}(r) = \langle [u(r) - u(0)]^3 \rangle = -\frac{4}{5}\,\varepsilon\, r , \tag{2.97}$$

discussed by Monin and Yaglom (1975, p. 397), or Frisch (1995, p.76).

A model for the Eulerian covariance (Durbin 1980) reads:

$$R_{\mathrm{uu}}(r) = \langle u^2 \rangle \left[1 - \left(\frac{r^2}{r^2 + L_{\mathrm{uu}}^2} \right)^{1/3} \right] , \tag{2.98}$$

which gives an integral length scale L_{D} different from L_{uu}:

$$L_{\mathrm{D}} = \frac{1}{\langle u^2 \rangle} \int_0^\infty R_{\mathrm{uu}}(r)\, \mathrm{d}r \approx 0.75 L_{\mathrm{uu}} . \tag{2.99}$$

Note that, for $r \ll L_{\mathrm{uu}}$:

$$D_{\mathrm{uu}}(r) = 2[\langle u^2 \rangle - R_{\mathrm{uu}}(r)] \approx 2\langle u^2 \rangle \left(\frac{r}{L_{\mathrm{uu}}} \right)^{2/3} , \tag{2.100}$$

which is K41 compliant. Comparing with (2.96), an Eulerian expression for the dissipation results:

$$\varepsilon = \frac{2^{3/2}\, \langle u^2 \rangle^{3/2}}{C_{\mathrm{K}}^{3/2} L_{\mathrm{uu}}} \tag{2.101}$$

From the definitions (2.56) and (2.95):

$$D_{\mathrm{uu}}(r) = 2 \int_0^\infty [1 - \cos(kr)]\, E_{\mathrm{uu}}(k)\, \mathrm{d}k , \tag{2.102}$$

so that given the Eulerian structure function (2.96), the spectrum in the inertial subrange reads:

$$E_{\mathrm{uu}}(k) = 0.249 C_{\mathrm{K}} \varepsilon^{2/3} k^{-5/3} , \tag{2.103}$$

(see Exercise 2.7). Note that this expression holds for wavenumbers smaller than $\mathrm{k_L} \approx L_{\mathrm{uu}}^{-1}$, corresponding to the integral scale, and larger than $\mathrm{k_K} \approx \mathrm{l_K^{-1}}$, corresponding to the Kolmogorov scale (2.92).

2.5.1.2 A Model Spectrum, from Olesen et al. (1984)

Model spectra have been discussed by Olesen et al. (1984). The easiest-to-integrate form is presented in Exercise 2.8 and reads:

$$E_{\mathrm{uu}}(k) = \frac{C_{\varepsilon} \varepsilon^{2/3}}{(b^{-1} + k)^{5/3}} , \tag{2.104}$$

where b is a length scale related to the velocity variance and to dissipation:

$$b \approx 1.55 \frac{\langle u^2 \rangle^{3/2}}{\varepsilon} . \tag{2.105}$$

2.5.1.3 From k to ω: Spectra in the Frozen Turbulence Approximation

The equivalence of the spectrum in frequency (ω) obtained from fixed point time series of wind velocity and that expressed in wavenumber (k) follows from the assumption that a mean velocity U can be defined.

Given:

$$k = 2\pi \frac{\omega}{U} , \tag{2.106}$$

it results:

$$E_{\mathrm{uu}}(k)\, \mathrm{d}k = E_{\mathrm{uu}} \left(\frac{2\pi\omega}{U} \right) \frac{2\pi}{U} \mathrm{d}\omega , \tag{2.107}$$

equivalent to, from (2.103),

$$E_{\mathrm{uu}}(k)\, \mathrm{d}k = \frac{0.249\, C_{\mathrm{K}}}{(2\pi)^{2/3}} U^{2/3} \varepsilon^{2/3} \omega^{-5/3}\, \mathrm{d}\omega . \tag{2.108}$$

We define $E_{uu}(\omega)$ as:

$$E_{uu}(\omega) = C_1 U^{2/3} \varepsilon^{2/3} \omega^{-5/3} , \tag{2.109}$$

with $C_1 = 0.249\, C_K/(2\pi)^{2/3}$. Note that this is an Eulerian spectrum.

2.5.1.4 Lagrangian Description

Along a trajectory:

$$D_{uu}(t) = \langle [v(t) - v(0)]^2 \rangle = 2\left(\langle u^2 \rangle - R_{uu}(t)\right) , \tag{2.110}$$

where $\langle v(\tau)^2 \rangle = \langle u^2 \rangle$ for any τ (thanks to homogeneity and steadiness). In the inertial subrange the second-order structure must be a function only of the dissipation and the time:

$$D_{uu}(t) \approx C_0 \varepsilon t , \tag{2.111}$$

where $C_0 = 6.2$ (Ouellette et al. 2006).

A model for the Lagrangian covariance (Gifford 1982) is:

$$R_{uu}(t) = \langle u^2 \rangle \exp(-t/T_{uu}) . \tag{2.112}$$

For $t \ll T_{uu}$, in the inertial subrange, $R_{uu}(t) \approx \langle u^2 \rangle (1 - t/T_{uu})$ and $D_{uu}(t) \approx 2\langle u^2 \rangle t/T_{uu}$. Comparing with (2.111) we obtain an expression for the dissipation as a function of the integral quantities characterizing the flow (Tennekes 1982):

$$\varepsilon = \frac{2\langle u^2 \rangle}{C_0 T_{uu}} . \tag{2.113}$$

The Lagrangian spectrum results, from (2.55) and (2.112):

$$E_{uu}(\omega) = \frac{2\langle u^2 \rangle T_{uu}}{\pi} \frac{1}{1 + \omega^2 T_{uu}^2} , \tag{2.114}$$

which, for $\omega \gg T_{uu}^{-1}$, gives the spectral slope in the inertial subrange $E(\omega) \propto \omega^{-2}$.

2.5.1.5 The Eulerian-to-Lagrangian Time Scale Ratio

By eliminating ε from Eqs. (2.113) and (2.101) a relationship among the integral scales characterizing the inertial sublayer is obtained:

$$\frac{L_{\mathrm{uu}}}{T_{\mathrm{uu}} \langle u^2 \rangle^{1/2}} = \frac{2^{1/2}\, C_0}{C_{\mathrm{K}}^{3/2}} \tag{2.115}$$

which gives an estimate of the Eulerian-to-Lagrangian characteristic time scale ratio (Maurizi et al. 2004).

2.5.2 *The Spectra of the Passive Tracer Variances*

The dimensional argument that leads to the Eulerian energy spectrum in the inertial subrange (2.103), can be used to find the shape of the spectrum of a passive scalar, including also the temperature (see Monin and Yaglom 1975, pp. 377–385). The spectrum must be function of the scalar dissipation rate (2.77) and of the TKE dissipation rate (2.74). Thus:

$$E_\theta(k) = C_\theta \varepsilon_\theta \varepsilon^{-1/3} k^{-5/3} \; . \tag{2.116}$$

Note that both the dissipation rates appear in the expression of the spectrum; the slope is the same as for the energy spectrum.

2.5.3 *Some Consequences of Isotropy*

Isotropy means that all the vectors must be null, as well as odd order single-point tensors. Moreover, the velocity variance-covariance tensor reduces to $\langle u_i u_j \rangle = \alpha \delta_{ij}$, i.e., the variances of the velocity components must be equal, and the off-diagonal terms null (Wyngaard 2010, p. 316).

In isotropic incompressible conditions and at high Reynolds number, the following results are relevant for the present study:

- the turbulence-mean flow coupling terms in (2.70) are null:

$$-\langle u_i u_k \rangle \frac{\partial U_i}{\partial x_k} = -\frac{\langle u_j u_j \rangle}{3} \delta_{ik} \frac{\partial U_i}{\partial x_k} = -\frac{\langle u_j u_j \rangle}{3} \frac{\partial U_i}{\partial x_i} = 0 \; , \tag{2.117}$$

because of the continuity equation. Also the heat flux terms are null, so that no turbulence production exists in isotropic conditions;

- the dissipation term in (2.70) can be written as:

$$2\nu\left\langle \frac{\partial u_i}{\partial x_j}\frac{\partial u_k}{\partial x_j}\right\rangle = \delta_{ik}10\nu\left\langle\left(\frac{\partial u_1}{\partial x_1}\right)^2\right\rangle ; \qquad (2.118)$$

- similarly, from (2.74):

$$\varepsilon = \nu\left\langle \frac{\partial u_i}{\partial x_j}\frac{\partial u_i}{\partial x_j}\right\rangle = 15\nu\left\langle\left(\frac{\partial u_1}{\partial x_1}\right)^2\right\rangle ; \qquad (2.119)$$

- the dissipation of the heat fluxes, derived from (2.79) neglecting the Laplacians, is null in isotropic conditions.
- the dissipation for temperature (and for any scalar) variance in isotropic conditions, from (2.77), reads:

$$\varepsilon_\theta = 3\kappa_{\mathrm{H}}\left\langle\left(\frac{\partial \theta}{\partial x_1}\right)^2\right\rangle . \qquad (2.120)$$

2.5.4 Final Remarks

Hereinafter, when a generic indication about time or length scales is necessary, we shall use T for the Lagrangian integral time scale and L_{E} for the Eulerian integral length scale, neglecting any reference to direction (or component) because of the isotropy of the inertial subrange. With reference to (2.115), we also identify an Eulerian time scale, which in isotropic conditions reads:

$$T_{\mathrm{E}} = \frac{3L_{\mathrm{E}}}{2E_{\mathrm{k}}^{1/2}} . \qquad (2.121)$$

We must remember that in geophysical flows anisotropy is the rule, because of the effect of stratification. The K41 paradigm can be realized at the smallest scales, but sometimes the isotropy subrange disappears almost completely: this point will be considered in some detail in Sect. 3.8. However, K41 remains a basis for most of the parameterizations we use in the applications, and an almost unavoidable check for the data, even not always justified.

Exercises

2.1 Derive (2.8) from (2.5) in the hypothesis of small perturbations.

2.2 Derive the conservation equation (2.16) for the mixing ratio in a compressible fluid.

2.3 Derive (2.22) and show that the potential temperature is constant in an adiabatic atmosphere.

2.4 Show that the commutative property holds with respect to the time derivative.

2.5 Show how the time average applied to a scalar variable modifies the power spectrum.

2.6 Compute the double integral (2.50), being the integrand an even function.

2.7 Show the general integration rule for (2.102) with $D = Ar^{\gamma}$.

2.8 Determine the coefficients of the spectrum (2.104).

References

G.T. Csanady, *Turbulent Diffusion in the Environment* (D. Reidel Pu. Co., Dordrecht, 1973)

J.W. Deardorff, A numerical study of three-dimensional turbulent channel flow at large Reynolds numbers. J. Fluid Mech. **41**, 453–465 (1970)

P.A. Durbin, A stochastic model for two-particle dispersion and concentration fluctuations in homogeneous turbulence. J. Fluid Mech. **100**, 279–302 (1980)

J.J. Finnigan, R.H. Shaw, Double-averaging methodology and its application to turbulent flow in and above vegetation canopies. Acta Geophys. **56**, 534–561 (2008)

U. Frisch, *Turbulence* (Cambridge University Press, Cambridge, 1995), 296 p.

C.W. Gardiner, *Handbook of Stochastic Methods for Physics, Chemistry and the Natural Sciences*, 2nd edn. (Springer, Berlin, 1990)

F.A. Gifford, Horizontal diffusion in the atmosphere: a Lagrangian-dynamical theory. Atmos. Environ. **15**, 505–512 (1982)

A.E. Gill, *Atmosphere-Ocean Dynamics* (Academic Press, San Diego, 1982)

G. Heskestad, A generalized Taylor's hypothesis with application for high Reynolds number turbulent shear flows. J. Appl. Mech. **87**, 735–739 (1965)

C.W. Higgins, M. Froidevaux, V. Simeonov, N. Vercauteren, C. Barry, M. Parlange, The effect of scale on the applicability of Taylor's frozen turbulence hypothesis in the atmospheric boundary layer. Bound.-Layer Meteorol. **143**, 379–391 (2012)

A.N. Kolmogorov, The local structure of turbulence in incompressible viscous fluid for very large Reynolds numbers. Dokl. Akad. Nauk SSSR **30**, 301 (1941)

A. Maurizi, G. Pagnini, F. Tampieri, Influence of Eulerian and Lagrangian scales on the relative dispersion properties in Lagrangian stochastic models of turbulence. Phys. Rev. E **69**, 037301–1/4 (2004)

A.S. Monin, A.M. Yaglom, *Statistical Fluid Mechanics*, vol. I (MIT Press, Cambridge, 1971), 769 pp.

A.S. Monin, A.M. Yaglom, *Statistical Fluid Mechanics*, vol. II (MIT Press, Cambridge, 1975), 874 pp.

H.R. Olesen, S.E. Larsen, J. Hojstrup, Modelling velocity spectra in the lower part of the planetary boundary layer. Bound.-Layer Meteorol. **29**, 285–312 (1984)

N.T. Ouellette, H.T. Xu, M. Bourgoin, E. Bodenschatz, Small-scale anisotropy in Lagrangian turbulence. New J. Phys. **8**, 1–10 (2006)

S. Pope, *Turbulent Flows* (Cambridge University Press, Cambridge, 2000)

E.A. Spiegel, G. Veronis, On the Boussinesq approximation for a compressible fluid. Astrophys. J. **131**, 442–447 (1960)

H. Tennekes, Similarity relations, scaling laws and spectral dynamics, *Atmospheric Turbulence and Air Pollution Modeling*, ed. by F.T.M. Nieuwstadt, H. van Dop (D. Reidel, Dordrecht, 1982), pp. 37–68

J.C. Wyngaard, *Turbulence in the Atmosphere* (Cambridge University Press, Cambridge, 2010)

J.C. Wyngaard, O.R. Coté, The budgets of turbulent kinetic energy and temperature variance in the atmospheric surface layer J. Atmos. Sci. **28**, 190–201 (1971)

J.C. Wyngaard, O.R. Coté, Y. Izumi, Local free convection, similarity, and the budgets of shear stress and heat flux. J. Atmos. Sci. **28**, 1171–1182 (1971)

Chapter 3
The Basic Paradigm: Horizontal Homogeneity Over Flat Terrain

Abstract The idealized conditions of horizontal homogeneity allow to develop the basic theory for the PBL. In this chapter we focus on what happens when the PBL processes depend only on the vertical coordinate and on the time. The similarity theory and the data (mainly from the field) lead to the formulation of parameterizations which are the essential background for the understanding of the observations and for the numerical modelling.

3.1 The Governing Equations

From Eqs. (2.68), (2.64) and (2.65) under conditions of horizontal homogeneity of the dynamic fields (i.e. $\partial/\partial x = \partial/\partial y = 0$) and flat horizontal terrain, it results:

$$\frac{\partial U_i}{\partial t} = \varepsilon_{ij3} f \left(U_j - U_{gj}\right) + \frac{\partial}{\partial z}\left(\nu \frac{\partial U_i}{\partial z} - \langle u_i w \rangle\right), i = 1, 2 , \tag{3.1}$$

$$W(z) = 0 , \tag{3.2}$$

$$\frac{\partial \Theta}{\partial t} = \frac{\partial}{\partial z}\left(\kappa_{\mathrm{H}} \frac{\partial \Theta}{\partial z} - \langle \theta w \rangle\right) . \tag{3.3}$$

Equation (3.1) can be written in terms of the complex variables $\mathbf{V} = U + \imath V$, $\mathbf{V}_{\mathrm{g}} = U_{\mathrm{g}} + \imath V_{\mathrm{g}}$ and $\mathbf{X} = \langle uw \rangle + \imath \langle vw \rangle$, representing the horizontal components of the mean wind, the geostrophic wind and the vertical flux of momentum:

$$\frac{\partial \mathbf{V}}{\partial t} = -\imath f(\mathbf{V} - \mathbf{V}_{\mathrm{g}}) + \nu \frac{\partial^2 \mathbf{V}}{\partial z^2} - \frac{\partial \mathbf{X}}{\partial z} . \tag{3.4}$$

The TKE equation, from (2.73), using the complex variables, reads:

$$\frac{\partial E_{\mathrm{k}}}{\partial t} = \mathscr{P} + \mathscr{B} - \frac{\partial \mathscr{T}}{\partial z} + \nu \frac{\partial^2 E_{\mathrm{k}}}{\partial^2 z} - \varepsilon , \tag{3.5}$$

where

$$\mathscr{P} = -\mathbf{X} \cdot \frac{\partial \mathbf{V}}{\partial z} \tag{3.6}$$

F. Tampieri, *Turbulence and Dispersion in the Planetary Boundary Layer*, Physics of Earth and Space Environments, DOI 10.1007/978-3-319-43604-3_3

is the shear production term,

$$\mathscr{B} = \frac{g}{\theta_{00}} \langle \theta w \rangle \tag{3.7}$$

is the buoyancy term, and

$$\mathscr{T} = \left[\frac{\langle p w \rangle}{\rho_{00}} + \frac{1}{2} \langle u_i u_i w \rangle \right], i = 1, 2, 3 \tag{3.8}$$

are the third-order terms.

The buoyancy term (3.7) may be positive or negative depending on the sign of the heat flux: positive heat flux (from the surface to the atmosphere) corresponds to production of TKE, negative heat flux to destruction. The shear production term (3.6) is positive if the mean shear is positive, as occurs near the ground (Sect. 2.4.6), so it corresponds to the production of TKE. The balance between these two terms is described by the flux Richardson number R_{f}, that reads:

$$R_{\mathrm{f}} = \left(\frac{g}{\theta_{00}} \langle \theta w \rangle \right) \left(\mathbf{X} \cdot \frac{\partial \mathbf{V}}{\partial z} \right)^{-1} . \tag{3.9}$$

Note that the flux Richardson number is a local variable. $R_{\mathrm{f}} < 0$ means that the heat flux feeds turbulence (unstable cases), while $R_{\mathrm{f}} > 0$ corresponds to the heat flux that decreases turbulence (stable cases). $R_{\mathrm{f}} = 0$ corresponds to the (ideal) neutral case: more generally, $|R_{\mathrm{f}}| \ll 1$ indicate that the heat flux is negligible in the balance of TKE.

Equation (3.5) may be written as:

$$\frac{\partial E_{\mathrm{k}}}{\partial t} = -\mathbf{X} \cdot \frac{\partial \mathbf{V}}{\partial z} (1 - R_{\mathrm{f}}) - \frac{\partial \mathscr{T}}{\partial z} + \nu \frac{\partial^2 E_{\mathrm{k}}}{\partial^2 z} - \varepsilon , \tag{3.10}$$

and it suggests that, neglecting the divergence of the third-order terms, for $R_{\mathrm{f}} > 1$ the production of turbulence due to the shear cannot balance the dissipation, so that turbulence cannot be steady. Accordingly, a critical value R_{fc} for the Richardson number in stable conditions can be defined: an empirical estimate is $0.20 \div 0.25$ (Grachev et al. 2013; Zilitinkevich et al. 2013). Below such value, steady conditions can occur. Above such value, intermittent or decaying turbulence is expected to occur, depending on the forcing mechanisms.

The observations suggest that the Richardson number can reach values larger than the critical one in stably-stratified turbulent flows. The TKE balance is not enough to deal with such conditions, and the equation for the mean turbulent potential energy (TPE) E_{P} must be considered. The TPE represents the potential energy associated with the vertical (adiabatic) displacements of fluid parcels in an environment characterized by a stable density gradient. The variation of the turbulent potential energy per unit mass, for a parcel undergoing a small vertical displacement l, reads:

$$\delta E_P = -\frac{g}{\rho_{00}} \int_0^l \tilde{\rho}\, dz = \frac{g}{\theta_{00}} \int_0^l \tilde{\theta}\, dz\ , \tag{3.11}$$

where (2.8) is used. By integrating and taking the average of the fluctuations (see Exercise 3.1), the TPE results:

$$E_P = \frac{1}{2} \frac{g^2}{\theta_{00}^2 N^2} \langle \theta^2 \rangle\ . \tag{3.12}$$

Accordingly, by multiplying (2.75) by $[g/(\theta_{00}N)]^2/2$ the equation for the TPE results (written in the present approximation, i.e. for conditions of horizontal homogeneity):

$$\frac{\partial E_P}{\partial t} = -\frac{g}{\theta_{00}} \langle \theta w \rangle + \frac{1}{2} \left(\frac{g}{\theta_{00} N} \right)^2 \frac{\partial \langle \theta^2 w \rangle}{\partial z} - \left(\frac{g}{\theta_{00} N} \right)^2 \varepsilon_\theta \tag{3.13}$$

where the Brunt-Väisälä frequency is now computed from the mean temperature profile: $N^2 = (g/\theta_{00})\, d\Theta_f/\, dz$ and is assumed to change slowly with respect to the TKE and TPE changes. The first term of RHS shows that the negative production term in the TKE equation is a positive term in the TPE equation: a negative heat flux reduces the TKE but increases the TPE, i.e. the temperature fluctuation variance. The interplay between TKE and TPE was suggested by Bolgiano (1959, 1962) and widely discussed by Zilitinkevich et al. (2008, 2013): see also Sect. 3.7 for some experimental observations.

Strictly steady conditions are compatible with the rotation of the wind, the momentum flux depending on z, the heat flux constant with z and the Richardson number smaller than the critical value.

3.2 Inner and Outer Scaling from the Wind Profile

At the steady state, (3.4) reads:

$$-\imath f(\mathbf{V} - \mathbf{V}_g) + \nu \frac{d^2 \mathbf{V}}{dz^2} - \frac{d\mathbf{X}}{dz} = 0\ . \tag{3.14}$$

As far as the PBL is driven by the surface-atmosphere interaction, the square root of the absolute value of the vertical momentum flux at the surface (the surface values are indicated by the subscript 0) gives a velocity scale, called friction velocity:

$$u_* = \left(\langle uw \rangle_0^2 + \langle vw \rangle_0^2 \right)^{1/4}\ . \tag{3.15}$$

Define $\tau = |\mathbf{X}|$, thus $u_*^2 = \tau_0$. In general, τ is expected to decrease with height.

The actual velocity departs from the geostrophic value because of the friction at the ground, so u_* can be used in (3.14) to scale the velocity difference (velocity defect) $\mathbf{V} - \mathbf{V}_\mathrm{g}$. Thus, neglecting molecular viscosity, a scale height of variation of the momentum flux can be defined as $u_*^2/(\mathrm{d}\tau/\mathrm{d}z) = (u_*/f)(u_*/|\mathbf{V} - \mathbf{V}_\mathrm{g}|)$, leading to an estimate of the PBL depth:

$$h = \alpha \frac{u_*}{f} , \tag{3.16}$$

with $\alpha \approx 0.2 \div 0.3$ derived from the observations (Garratt 1992). According to this formulation, the PBL depth (in absence of heat flux effects) is the result of the balance between momentum transfer and rotation effect; it does not work near the Equator and also under conditions of weak surface momentum flux and strong stratification of the lower atmosphere (large N) like in the very cold regions (see Zilitinkevich and Esau 2002, for a discussion). The argument is extended in Sect. 3.7.4.

Let us consider a vertical scale h_s much smaller than the PBL depth: $h_\mathrm{s} f/u_* \ll 1$: using h_s to scale the heights in (3.14) it results:

$$\frac{\mathrm{d}[(\tau + \tau_\nu)/u_*^2]}{\mathrm{d}(z/h_\mathrm{s})} \approx 0 , \tag{3.17}$$

where $\tau_\nu = \nu\, \mathrm{d}|\mathbf{V}|/\mathrm{d}z$ is the absolute value of the viscous momentum flux. Thus, in a layer of depth much smaller than the PBL depth, the momentum flux may be assumed to be constant with height: this layer is called surface layer, or constant-flux layer. In this layer the rotation is negligible (the relevant Rossby number $u_*/h_\mathrm{s} f$ is much larger than 1), so the mean wind lies in a plane: the x-axis may be oriented as the mean wind, U being its only non-zero component. For symmetry reasons $\langle vw \rangle = 0$.

3.3 Similarity, Obukhov Length and Beyond

The similarity approach means that physical arguments lead to find appropriate scales from the driving quantities in the surface layer, and then to formulate in nondimensional form the statistical moments of the turbulent quantities. This approach leads, if successfull, to individuate general dependencies among nondimensional variables; the resulting mathematical relations are characterized by numerical coefficient determined from the experimental observations.

A widely used approach is based on the assumption that surface turbulent fluxes drive the PBL dynamics and thus determine the relevant scales (Monin and Yaglom 1971): it is usually indicated as Monin Obukhov Similarity Theory (MOST). The straightforward application refers to the surface layer, with constant fluxes. The extension to cases of fluxes changing with height will be considered as appropriate.

We choose u_* as the velocity scale (Sect. 3.2). From the heat flux $\langle \theta w \rangle$ the temperature scale can be derived:

$$\theta_* = \langle \theta w \rangle_0 / u_* \, . \tag{3.18}$$

From these two scales (both pertaining to the vertical fluxes), a length scale results:

$$L = -\frac{\theta_{00} \, u_*^2}{\kappa \, g \, \theta_*} \, , \tag{3.19}$$

that was at first introduced by Obukhov (1946) and is called Obukhov length. In this definition the parameter κ is introduced: it is expected to be a universal constant, called the von Karman constant, and its value is derived from the observations. Note that the meaning of L does not change if κ is omitted from the definition.

The Obukhov length can be interpreted as a measure of the stability. The sign of the heat flux determines the sign of L, which is the same of the flux Richardson number: negative for unstable cases (positive heat flux), positive for stable cases. The absolute value of L decreases as the absolute value of the heat flux increases; for negligible heat flux (quasi-neutral cases), L diverges and cannot be used as a scale. Moreover, L can be used to normalize the height z above the surface: for z smaller than $|L|$ the effect of the heat flux on turbulence is expected to be negligible, while it is expected to be relevant for z larger than $|L|$. In other words, independently from the stability conditions, the heat flux has negligible effects near the ground where the momentum flux is dominant, while it becomes important as the height increases.

Stability can be measured also in terms of the gradients of mean velocity and temperature, using the gradient Richardson number, which is derived from the flux Richardson number (3.9) by substituting the fluxes with the corresponding relationships for the gradients (2.80) and (2.85), and dropping the ratio between the eddy diffusion coefficients. The gradient Richardson number reads:

$$R_g = \frac{g}{\Theta_r} \frac{d\Theta / dz}{(dU/dz)^2 + (dV/dz)^2} \, . \tag{3.20}$$

Here Θ_r is the mean temperature at the height where the Richarson number is evaluated. R_g has the same sign properties of R_f.

For practical instances, the bulk Richardson number R_B (the discrete version of R_g) can be defined:

$$R_B(z_1, z_2, z_3, z_4) = \frac{g}{\Theta_r} \frac{\Theta(z_2) - \Theta(z_1)}{[U(z_4) - U(z_3)]^2} \frac{[z_4 - z_3]^2}{z_2 - z_1} \, , \tag{3.21}$$

where the dependence on the heights at which the mean quantities are evaluated is explicitly indicated. R_B can be understood also as a bulk value for a finite thickness layer of air.

The similarity approach implies the use of the scales to make the variables nondimensional: it works as far as these scales are well determined from the data. Thus for negligible friction velocity and/or heat flux, the similarity becomes meaningless. Moreover, the experimental errors make the determination of these scales quite unreliable. A side problem in the use of scaled variables arises because of the possible self correlation between variables, as the same scale appears in the dependent and the independent one, and the scale is affected by errors. These points are discussed mainly in the frame of stable boundary layers, for instance by Baas et al. (2006), Klipp and Mahrt (2004), Grachev et al. (2007), Liang et al. (2014), but deserve special attention for any analysis based on scaled quantities: see for instance the discussion about Fig. 3.5.

Similarity approaches based on scales derived from quantities other than turbulent fluxes have been developed (Sorbjan 2010); they are not used in this book.

3.4 The Surface Layer in Neutral and Unstable Conditions

The surface layer is characterized by steady, horizontally homogeneous conditions and constant fluxes: Eqs. (3.1) and (3.3) together with (3.17) do not allow to compute a solution for the mean horizontal wind and temperature, so dimensional arguments and data analysis must be used.

The relevant velocity and temperature scales are u_* and θ_*, and the height scales are h_s and L. The boundary layer depth may affect the surface layer quantities: this effect is negligible in many cases, because $h_s \ll h$, but this inequality may be not satisfied if h is not given by (3.16), as can occur for instance in stable conditions.

3.4.1 The Quasi-Neutral Conditions and the Mean Wind Profile

We define the quasi-neutral conditions as $|z/L| \ll 1$: the flux Richardson number is negligible as well as the heat flux (and thus θ_*), the velocity scale is u_*, while a length scale cannot be identified from the fluxes. Note that the numerical value of the limit of $|z/L|$ for defining neutrality is matter of observations, and in practice it can be chosen on the basis of the sensitivity of the measurements. As a rule of thumb, a value of 0.01 is often appropriate.

Equation (3.17) states that the total momentum flux is constant with z. Using ν/u_* to make the height nondimensional, i.e. defining $z^+ = z u_*/\nu$, (3.14) becomes:

$$\frac{\tau}{u_*^2} + \frac{\mathrm{d}(U/u_*)}{\mathrm{d}z^+} = 1 \,. \tag{3.22}$$

Very near to the surface, i.e. for $z^+ < 1$, the variance of the vertical velocity fluctuations w is expected to become smaller and smaller because of the presence of the rigid boundary while the viscous term is of $\mathcal{O}(1)$. Thus the velocity profile is dominated by the viscous effects; the velocity increases linearly with height:

$$\frac{U(z)}{u_*} = z^+ . \tag{3.23}$$

The Reynolds number based on the distance from the wall increases as this distance increases, so that the viscous term may be neglected and the momentum flux is due to turbulence only: $\langle uw \rangle$ is constant and equal to u_*^2. Note that this observation explains the value of 1 at RHS in (3.22). The gradient of the mean velocity must depend on the friction velocity and on the distance from the wall only:

$$\frac{\mathrm{d}U}{\mathrm{d}z} = \frac{u_*}{\kappa z} , \tag{3.24}$$

where the proportionality constant is the von Karman constant, already used in (3.19), whose value is normally determined from the analysis of wind profiles, and in this book will be assumed to be 0.4. A recent discussion about its variability can be found in Andreas et al. (2006).

The von Karman constant can be interpreted as a numerical constant that enters in the eddy diffusion coefficient. In terms of the flux-gradient relationship, using (3.24) in (2.84) and (2.80), it results that the eddy diffusion coefficient in a quasi-neutral surface layer is

$$K_\tau = \kappa u_* z , \tag{3.25}$$

and that the mixing length is proportional to the distance from the ground

$$l_\tau^2 \equiv \langle \alpha l^2 \rangle = \kappa^2 z^2 . \tag{3.26}$$

Equation (3.24) entails a logarithmic profile in the high Reynolds number region, i.e. for $z^+ \gg 1$:

$$U(z) - U(z_1) = \frac{u_*}{\kappa} \ln \frac{z}{z_1} . \tag{3.27}$$

From the observations, the matching with the linear profile occurs at $z_\mathrm{m}^+ \approx 11$ (Monin and Yaglom 1971) giving

$$U(z)/u_* = 1/\kappa \ln z^+ + \left(z_\mathrm{m}^+ - \ln z_\mathrm{m}^+/\kappa\right) \approx 2.5 \ln z^+ + 5 , z^+ > z_\mathrm{m}^+ , \tag{3.28}$$

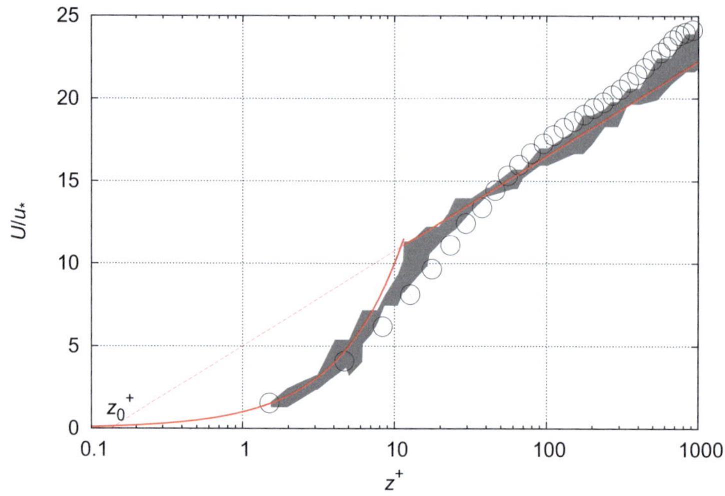

Fig. 3.1 Non-dimensional profile of the mean velocity as function of the non-dimensional height. The *gray* area, redrawn from Monin and Yaglom (1971), represents the envelope of the observations (mainly from laboratory measurements). The *open circles* refer to the numerical simulations by Uijttewaal and Oliemans (1996) at $R_e = 2100$. The *continuous lines* represent the linear profile (3.23) and the logarithmic one (3.28). The *dashed line* is the extrapolation of the logarithmic profile to $U = 0$, that identifies the value of the roughness length

which is the profile over a smooth surface in the turbulence-dominated layer. A surface is smooth if its irregularities have a representative (mean) height h_c smaller than $z_m^+ \nu/u_*$. It is worth noting that the concept of smoothness is related not only to the surface geometric characteristics, but also to the dynamics of the turbulent flow (to the transfer of momentum).

The data suggest that the logarithmic form can be used for $z^+ > 30$ and the linear one for $z^+ < 5$: see Fig. 3.1. A smooth matching is appropriate for intermediate values of z^+.

If the surface is characterized by obstacles of height $h_c \gg z_m^+ \nu/u_*$ (a rough surface), the viscous profile cannot be observed, although a layer affected by viscosity exists around the obstacles. In this case (which is the most frequent in the atmospheric boundary layer), only the logarithmic profile is observed, leading to define the roughness length z_0 as the height at which the extrapolated logarithmic profile goes to zero:

$$U(z) = \frac{u_*}{\kappa} \ln \frac{z}{z_0} \,. \tag{3.29}$$

The roughness length is essentially a dynamical quantity. Over a smooth surface it is a function of the viscosity and the friction velocity: from (3.28) it results

$$z_0 = z_m^+ \frac{\nu}{u_*} \exp\left(-\kappa z_m^+\right) \approx 0.135 \frac{\nu}{u_*} \,. \tag{3.30}$$

Over a rough surface, from (3.29), it results:

$$z_0 = z \exp\left(-\frac{\kappa}{\sqrt{c_U(z)}}\right), \tag{3.31}$$

where c_U is the drag coefficient:

$$c_U(z) = \frac{u_*^2}{U^2(z)}, \tag{3.32}$$

which gives a measure of the transfer of momentum from the atmosphere to the surface: small values of c_U indicate small vertical transfer of momentum. From (3.31) it results that the roughness length increases as the drag increases, in neutral conditions (in the shear-dominated surface layer).

The roughness length and the drag for a site covered by grass is reported in Fig. 3.2, as function of the mean wind velocity. The increase of z_0 as the wind reduces stresses the dynamic nature of this quantity, which reaches a constant value only for sufficiently high winds. It is also interesting to observe that the determinations of z_0 made from the drag at two different levels (Fig. 3.2a, black and gray symbols) are slightly different: this corresponds to the different overall conditions (season, state of the ground) being the site the same.

The roughness length is affected, but not uniquely determined, by the shape and the dimensions of the roughness elements (the obstacles) present on the

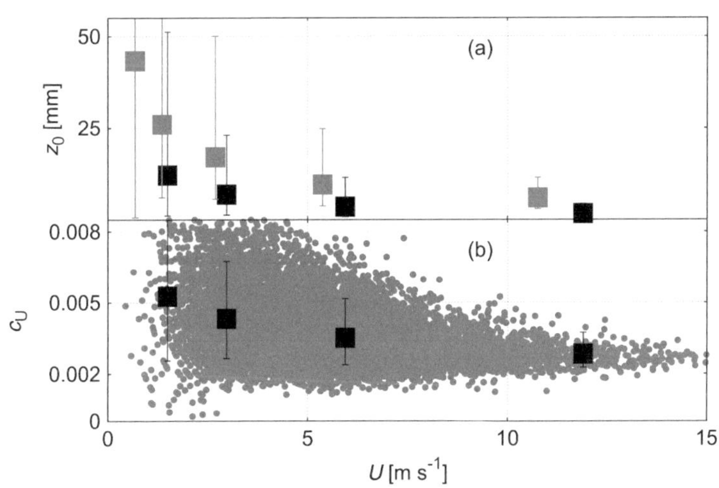

Fig. 3.2 Roughness length (**a**) and drag coefficient (**b**) over a grass surface as a function of wind speed measured at 10 m: *black squares*, and at 3 m: *gray squares*, in panel (**a**) only. *Gray dots* in (**b**) represent the full set of data. Observations from CABAUW dataset

surface. In most modeling applications, like the weather forecasts, or the air quality evaluations, the features of the surface (and thus the obstacle height) are known from independent informations, like land-use databases. Thus, as a rule of thumb, z_0 is taken as a fraction of h_c, ≈ 0.03 following Monin and Yaglom (1971, p. 289) for instance, or by means of tables detailing the kind of surface (see, for instance Garratt 1992, Table A6).

The presence of the obstacles can prevent the flow to penetrate downward to the surface $z = 0$, related to size and distribution of the obstacles themselves. Thus it may be convenient to displace the logarithmic profile of a quantity z_d, the displacement height, in order to get rid of the flow inside the canopy. Equation (3.29) then reads:

$$U(z) = \frac{u_*}{\kappa} \ln \frac{z - z_d}{z_0} , \tag{3.33}$$

and z_d is a dynamical quantity, of the order of the average height h_c of the obstacles.

On the sea, the roughness length depends on the wind, being related to the surface waves. The empirical formula by Charnock (1955) reads:

$$z_0 = \alpha u_*^2 / g , \tag{3.34}$$

where $0.01 < \alpha < 0.03$ (see Garratt 1992; Makin 2005), corresponding to a drag coefficient increasing with the wind. The measurements by Powell et al. (2003) show that over a threshold (broadly, for $U > 30\,\mathrm{m\,s^{-1}}$ at the standard measuring height of 10 m), the drag coefficient decreases: see Fig. 3.3. This effect is attributed to the presence of spray over the waves, and stresses the dynamic nature of the roughness length.

Quite obviously, the argument that leads to (3.24) and (3.27), applied to the mean temperature suggests that Θ is constant in the quasi-neutral surface layer, because the heat flux is close to zero as well as the temperature scale.

The friction velocity is also the scale for the velocity fluctuations, so that the ratios $\langle u_i^2 \rangle / u_*^2$ are expected to be constant (independent on the external conditions), following the same argument that leads to (3.24). This point will be discussed later (Sect. 3.4.3). In the atmospheric flows the temperature fluctuations are always present, due to departures from the ideal conditions of homogeneity and steadiness (see, for instance Tampieri et al. 2009). This aspect will be analyzed in Sect. 3.4.3.4.

3.4.2 *Unstable Conditions*

As the value of $\zeta = -z/L$ increases, the heat flux becomes more and more important; in conditions of intense convection, it is expected to determine the PBL

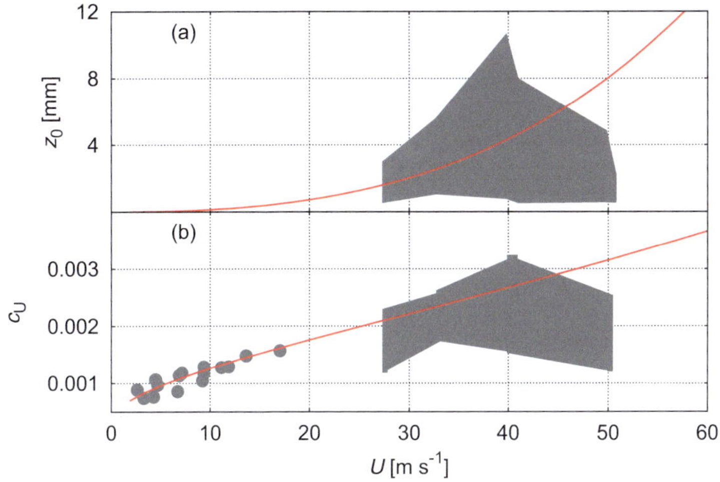

Fig. 3.3 Roughness length (**a**) and drag coefficient (**b**) over the sea as a function of wind speed measured at 10 m above the mean sea level. *Gray areas* represent the data envelope from Powell et al. (2003). *Gray dots* correspond to observations by Caulliez et al. (2008). The *red line* in (**a**) corresponds to (3.34) with $\alpha = 0.01$; in (**b**) to (3.32)

dynamics. In other words, as $\zeta > 1$, the friction velocity is not suited as velocity scale.

According to the analysis by Kader and Yaglom (1990), in the unstable surface layer, three sublayers may be devised, whose extension in terms of ζ is given here on the basis of the observations:

- a dynamic (quasi-neutral, shear-driven) sublayer, for $\zeta < 0.01$, where u_* is the scale for the velocity and θ_* is the scale for the temperature;
- a free-convective sublayer, for $\zeta > 1$, where the velocity scale is function of the heat flux only, and is defined as:

$$w_*(z) = \left(\frac{g}{\theta_{00}} \langle \theta w \rangle_0 \, z \right)^{1/3} . \tag{3.35}$$

A consistent choice of the temperature scale turns out to be:

$$\theta_{**}(z) = \frac{\langle \theta w \rangle_0}{w_*} = \frac{\langle w\theta \rangle_0^{2/3}}{[(g/\theta_{00})z]^{1/3}} . \tag{3.36}$$

Note that the velocity scale increases and the temperature scale decreases with height;
- an intermediate (transition) sublayer, for $0.1 < \zeta < 0.5$, characterized by two velocity scales: u_{**} specific for the horizontal components, and w_* for the vertical component, while the temperature scale is still θ_{**} (because it derives

from the vertical heat flux). The definition of u_{**} is:

$$u_{**}(z) = \frac{u_*^2}{w_*} \propto u_* \zeta^{-1/3} . \tag{3.37}$$

In this sublayer the velocity scale for the horizontal components decreases with height.

It is relevant to note here that the large convective eddies, and more generally the structure of the entire convective boundary layer affect the surface layer turbulence. This issue will be raised in the next Sections as appropriate: some general discussion can be found in Yaglom (1994) and in Zilitinkevich et al. (2006).

3.4.2.1 The Mean Velocity Profile

According to this three-layer scheme, the similarity relationship (3.24) is modified using the proper velocity scale in each layer.

We define the non-dimensional mean velocity gradient Φ_U:

$$\Phi_\mathrm{U} = \frac{\kappa z}{u_*} \frac{\mathrm{d}U}{\mathrm{d}z} . \tag{3.38}$$

In the dynamic sublayer, the velocity scale is u_*, and the result (3.24) for neutral conditions is recovered:

$$\Phi_\mathrm{U} = 1 . \tag{3.39}$$

In the intermediate sublayer, using u_{**} as scale velocity, it results:

$$\Phi_\mathrm{U} = \alpha_1 \zeta^{-1/3} . \tag{3.40}$$

In the free-convective sublayer the velocity scale is w_* and Φ_U should read:

$$\Phi_\mathrm{U} = \alpha_2 \zeta^{1/3} . \tag{3.41}$$

The numerical values of α_1 and α_2 are obtained from the observations (see Table 3.1).

The behaviour described by (3.41) is not found in the data analysis that follows (a part an historical exception, see Fig. 3.4) and in the literature (see, for instance Li et al. 2012; Hicks et al. 2014). The horizontal wind speed is affected by the shear near the surface (in the dynamic sublayer), and its gradient decreases by effect of convective motions in the transition between shear and free convection. The free-convection sublayer is thus characterized by w_* as a velocity scale, but it affects only marginally the horizontal components, while is important for the vertical one, as will be seen in the discussion about second-order moments.

Table 3.1 Numerical coefficients for mean wind formulas (unstable conditions)

Wyngaard et al. (1971)	$\alpha_1 = 15$	$\gamma = 1/4$	
Kader and Yaglom (1990)	$\alpha_1 = 0.48$	$\alpha_2 = 0.37$	
Högström (1996)	$\alpha_1 = 19$	$\gamma = 1/4$	$\zeta_1 = 2$
Frenzen and Vogel (2001)	$\alpha_1 = 16$	$\gamma = 1/3$	$\zeta_1 = 0.5$

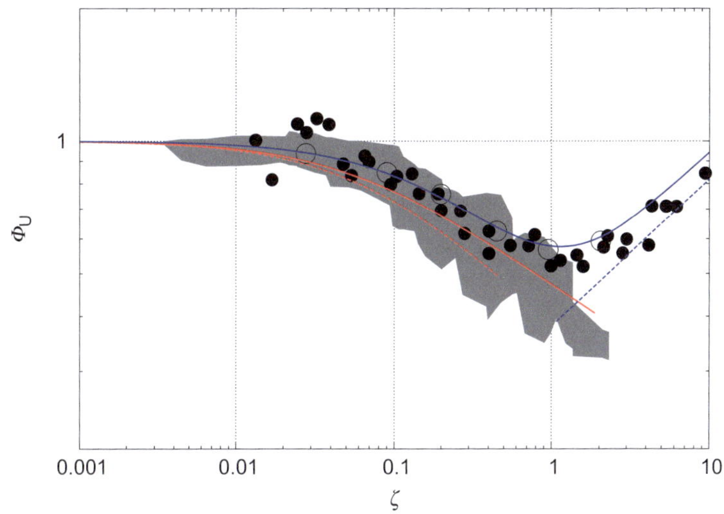

Fig. 3.4 Φ_U as function of ζ. The *gray area* represents the envelope of data from Högström (1996). The *open circles* are the data from Kader and Yaglom (1990) as reported by Högström (1996), while the *full dots* are all the data cited by Kader and Yaglom (1990). Note that Högström (1996) reported only the first point supporting the increase of Φ_U. The *red lines* correspond to (3.42) with the parameters suggested by Högström (1996) (*full line*) and those suggested by Frenzen and Vogel (2001) (*dashed line*); the *full blue line* to (3.43) and the *dashed blue line* to (3.41)

The formulations suggested by Wyngaard et al. (1971), Högström (1996) and Frenzen and Vogel (2001) read:

$$\Phi_U = (1 + \alpha_1 \zeta)^{-\gamma} \ , \text{ for } \zeta < \zeta_1 \tag{3.42}$$

where the values of the coefficients are given in Table 3.1. Note that the value of the exponent grounds on empirical bases for the two former authors. The non-dimensional velocity gradient suggested by Kader and Yaglom (1990) as an interpolation formula:

$$\Phi_U = \left(\frac{1 + 0.65\zeta^2}{1 + 7.5\zeta} \right)^{1/3} \tag{3.43}$$

is cited here for historical reasons. The non-dimensional velocity gradient from the observations reported by Högström (1996) and by Kader and Yaglom (1990) and the previous formulas are shown in Fig. 3.4. The data for large ζ are suspected by self-correlation: in fact, the variable in ordinate is $\propto u_*^{-1}$ while the abscissa contains u_*^{-3}: a positive error on the friction velocity leads to a reduction of the abscissa and of the ordinate too, leading to an artificial positive correlation.

The mean velocity profile is obtained by integration:

$$U(z_2) - U(z_1) = \frac{u_*}{\kappa} \int_{z_1}^{z_2} \frac{\Phi_U(\zeta)}{z} \, dz . \tag{3.44}$$

Integrating numerically (3.43) or using (3.39)–(3.41) (see Exercise 3.2), an expression consistent with the Kader and Yaglom (1990) scheme can be obtained. Beljaars and Holtslag (1991) suggest the following form:

$$\begin{aligned}\kappa \frac{U(z_2) - U(z_1)}{u_*} &= \ln \frac{z_2}{z_1} + [\arctan(\xi_2) - \arctan(\xi_1)] \\ &- \ln \left[\frac{(1+\xi_2)^2(1+\xi_2^2)}{(1+\xi_1)^2(1+\xi_1^2)} \right] ,\end{aligned} \tag{3.45}$$

where

$$\xi = (1 + 16\zeta)^{1/4} . \tag{3.46}$$

This form is commonly used in many applications and shows the decreasing behaviour typical of the intermediate sublayer. The profile consistent with Frenzen and Vogel (2001) is derived in Exercise 3.3.

The comparison with the data is presented in Fig. 3.5. In order to evidence the self-correlation problem, the data are reported for the full experimental range of u_* and filtered for large friction velocity (small experimental error). This choice reduces (but not eliminates) the impact of self-correlation. In spite of large scatter, the data are consistent with the prediction for the shear and intermediate layers: the normalized velocity difference decreases as the instability increases, while the increasing trend for large values of $-1/L$ is a manifestation of the self-correlation.

Note that the variability of the coefficients must be related to the departure from the ideal conditions: we shall face with the same problem in the following Sections, and we will discuss it again in Chap. 4.

3.4.2.2 The Mean Temperature Profile

As for the mean velocity, we define the non-dimensional temperature gradient Φ_Θ:

$$\Phi_\Theta = -\frac{\kappa z}{\theta_*} \frac{d\Theta}{dz} . \tag{3.47}$$

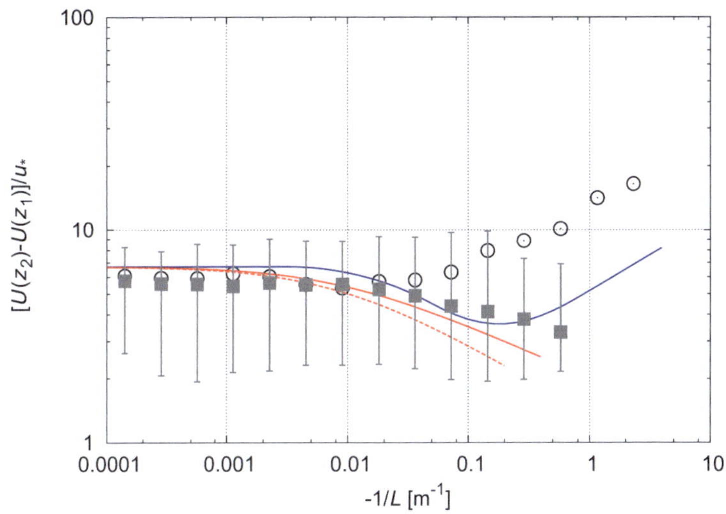

Fig. 3.5 $[U(z_2) - U(z_1)]/u_*$ as function of $-1/L$. CCT data (*full squares*: filtered with $u_* > 0.1\,\mathrm{m\,s^{-1}}$, *open circles*: unfiltered), $z_1 = 2\,\mathrm{m}$, $z_2 = 33.4\,\mathrm{m}$. The *full red line* corresponds to (3.45), the *dashed red line* to the integral of (3.42) with the parameters by Frenzen and Vogel (2001) (see Exercise 3.3), the *full blue line* to the integral of (3.39), (3.40) and (3.41) (see Exercise 3.2) with the coefficients reported in Table 3.1

Here the minus sign indicates that for positive heat flux (and thus positive θ_*), the mean temperature decreases with height. We identify two temperature scales, namely θ_* for the dynamic sublayer, and θ_{**} for the transition and free-convective sublayers. Thus we shall have two forms:

$$\Phi_\Theta = \alpha_0 \ , \tag{3.48}$$

in the dynamic sublayer, and

$$\Phi_\Theta = \alpha_1 \zeta^{-1/3} \ , \tag{3.49}$$

in the intermediate and in the free-convective sublayers. The values of the parameters are reported in Table 3.2. Kader and Yaglom (1990) suggest two slightly different values of the coefficient α_1 in the two sublayers (this distinction is neglected here) and an interpolation formula

$$\Phi_\Theta = 0.64 \left(\frac{3 + 2.5\zeta}{1 + 10\zeta + 50\zeta^2} \right)^{1/3} \ . \tag{3.50}$$

Table 3.2 Numerical coefficients for mean temperature formulas (unstable conditions)

Wyngaard et al. (1971)	$\alpha_0 = 0.74$	$\alpha_1 = 9$	
Kader and Yaglom (1990)	$\alpha_0 = 0.92$	$\alpha_1 = 0.28$	
Högström (1996)	$\alpha_0 = 0.95$	$\alpha_1 = 11.6$	$\zeta_1 = 2$

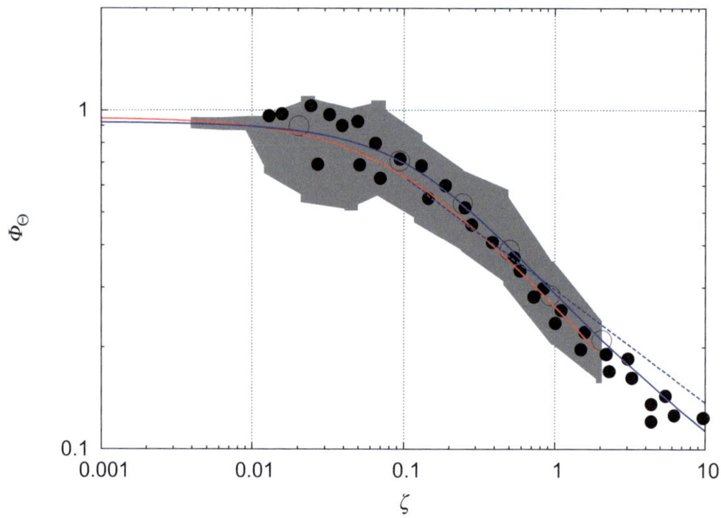

Fig. 3.6 Φ_Θ as function of ζ. *Gray area and circles* as in Fig. 3.4. The *full red line* corresponds to (3.51) with the parameters by Högström (1996), the *full blue line* to (3.50) and the *dashed blue line* to (3.49)

Wyngaard et al. (1971) and Högström (1996) recommend an empirical power law:

$$\Phi_\Theta = \alpha_0 \left(1 + \alpha_1 \zeta\right)^{-1/2} \text{ for } \zeta < \zeta_1 \; . \tag{3.51}$$

Some non-dimensional temperature gradient data and model curves are reported in Fig. 3.6. The numerical values of the coefficients used here are reported in Table 3.2.

The mean temperature profile is obtained by integration of (3.47). From (3.51) it results (Beljaars and Holtslag 1991):

$$\kappa \frac{\Theta(z_2) - \Theta(z_1)}{-\theta_*} = \ln \frac{z_2}{z_1} - 2 \ln \left(\frac{1 + \xi_2^2}{1 + \xi_1^2} \right) \tag{3.52}$$

where ξ is defined by (3.46). An expression consistent with Kader and Yaglom (1990) scheme may be obtained by integrating numerically (3.50) or using (3.48) and (3.49): see Exercise 3.2.

A comparison with some data is shown in Fig. 3.7.

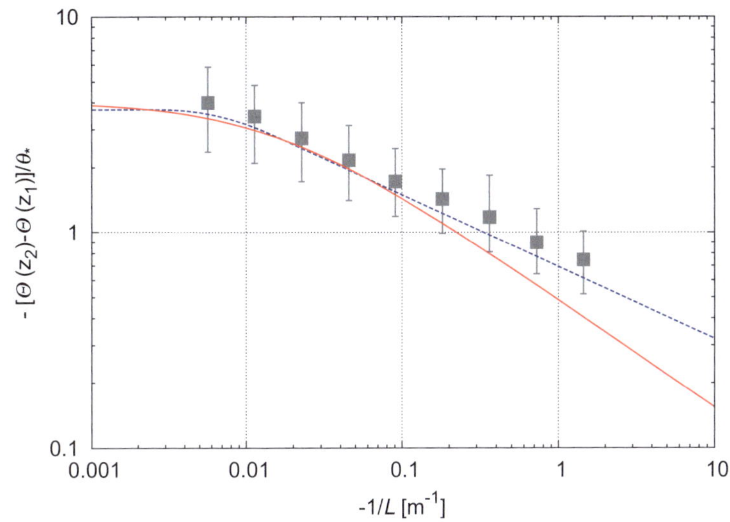

Fig. 3.7 $-[\Theta(z_2) - \Theta(z_1)]/\theta_*$ as function of $-1/L$. *Gray symbols*: CABAUW data, with the condition $\langle \theta w \rangle > 0.04\,\mathrm{K\,m\,s^{-1}}$ to avoid divergence at small ζ. Here $z_1 = 2\,\mathrm{m}$, $z_2 = 10\,\mathrm{m}$. The *full red line* corresponds to (3.52), the *dashed blue line* to the integration of (3.48) and (3.49): see (S3.12)

3.4.2.3 Richardson Numbers, Eddy Diffusion Coefficients and the Turbulent Prandtl Number

In the surface layer, the eddy diffusion coefficients for momentum and heat can be derived using (2.80) and (2.85). It results:

$$K_\tau = \kappa\, u_*\, z\, \Phi_{\mathrm{U}}^{-1} \ , \tag{3.53}$$

and

$$K_{\mathrm{Q}} = \kappa\, u_*\, z\, \Phi_{\Theta}^{-1} \ , \tag{3.54}$$

extending the quasi-neutral expression (3.25). These relations evidence that the eddy diffusion coefficients are functions of stability.

The gradient Richardson number, the flux Richardson number and the Obukhov length are related; in particular from (3.20):

$$R_{\mathrm{g}} = \zeta \frac{\Phi_\Theta}{\Phi_{\mathrm{U}}^2} \ , \tag{3.55}$$

and from (3.9):

$$R_{\mathrm{f}} = \zeta \Phi_{\mathrm{U}}^{-1} \,. \tag{3.56}$$

These expressions can be computed using the nondimensional gradients discussed above, and must be tested against data. Also R_{B} can be computed using the profiles for wind and temperature. A plot is reported in Fig. 3.8.

The turbulent Prandtl number $P_{\mathrm{t}} = K_{\tau}/K_{\mathrm{Q}}$ is also a function of stability. According to the parameterizations discussed before, it is larger than 1 also in neutral conditions.

Comparing (3.53) and (2.84), the mixing length increases in unstable conditions with respect to the neutral value (3.26):

$$\langle \alpha l^2 \rangle^{1/2} = \kappa z \Phi_{\mathrm{U}}^{-1} \,. \tag{3.57}$$

We anticipate here that the CBL as a whole (not limiting to the surface layer) is a typical example of turbulent flow characterized by eddies of the same spatial scale of its depth. Mixing is no doubt non-diffusive, leading to countergradient (non-Fickian) transport (Deardorff 1966): this is a critical issue in many applications, and will be discussed in Chap. 6.

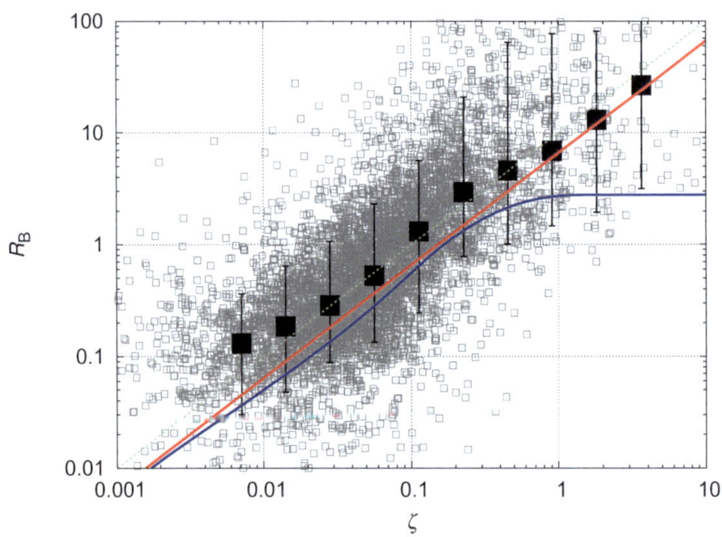

Fig. 3.8 R_{B} vs ζ from CABAUW data, estimated in the layer between 3 and 60 m (*gray symbols*: single observations; *black squares*: binned observations). ζ is computed from the data at 3 m. The *red line* is computed using the mean profiles from Beljaars and Holtslag (1991), the *blue line* from Kader and Yaglom (1990). The *green line* is the approximation $R_{\mathrm{B}} = 10\zeta$ used to drawn the abscissae in Fig. 3.10

3.4.3 *The Higher-Order Moments of the Velocity Components and of the Temperature Fluctuations*

Informations about the shape of the probability density function of the velocity and the temperature fluctuations can be obtained from the analysis of the statistical moments. Besides the knowledge of the atmospheric turbulence, these informations are relevant in the applications, for instance in the dispersion modelling. In this Section the second-order moments of velocity and temperature will be discussed with reference to neutral and unstable conditions. (Stable conditions are treated in Sect. 3.7.2.) As far the skewness and kurtosis are concerned, the few available data referring to the whole range of stability are presented.

We notice here that we shall face with an evident contradiction in the application of the similarity theory using coefficients tuned on the specific experiment, i.e. with the non-universality of the numerical coefficients. (The same problem we have faced in the previous Section looking at different data sets for the mean velocity.) The point is that the basic rules are quite robust and outline real physical dependencies; non-ideal effects (for instance, inhomogeneity, or unsteadiness) are not treated explicitly and give rise to the variability of some coefficients.

3.4.3.1 The Variances of the Velocity Components

Based on the same scaling rules applied to mean wind and temperature (Kader and Yaglom 1990), the variances of the velocity component fluctuations (and thus the TKE) scale on u_*^2 in the dynamic (quasi-neutral) sublayer and on w_*^2 in the free-convective sublayer (leading to $\langle u_i^2\rangle/u_*^2 \propto \zeta^{2/3}$). In the intermediate sublayer, the horizontal components are expected to scale on u_{**}^2, while the vertical one is expected to scale on w_*^2.

In conditions of L very small, the influence of the large eddies may become overwhelming, leading to dependencies on the boundary layer depth h instead of z. This point has been discussed by Yaglom (1994) and reviewed by Zilitinkevich et al. (2006).

The standard approach to find a universal (based on MOST) law describing the variances of the velocity components is as follows. Matching the scaling laws for the free-convective and quasi-neutral conditions, the following expression results:

$$\frac{\langle u_i^2\rangle}{u_*^2} = \alpha_i\,(1+\beta_i\zeta)^{2/3} \ , \tag{3.58}$$

where the values of the coefficients according to some authors are reported in Table 3.3. Remember that Kader and Yaglom (1990) use a different coefficient in the intermediate layer: it turns out to be 1.65, as shown in Fig. 3.9c. Note the different coefficients for different experiments, related to departures from ideal conditions, and to different analysis techniques. For instance, the filtering technique applied to

Table 3.3 Numerical coefficients for the variances of the velocity components (unstable conditions)

Kader and Yaglom (1990, quasi-neutral)	$\alpha_1 = 7.9$	
	$\alpha_2 = 6.3$	
	$\alpha_3 = 1.6$	
Kader and Yaglom (1990, free-convective)		$\beta_3 = 1.3$
Kaimal and Finnigan (1994)	$\alpha_3 = 1.6$	$\beta_3 = 3.0$
Andreas et al. (1998)	$\alpha_3 = 1.1$	$\beta_3 = 4.3$
de Franceschi et al. (2009, filtered, along valley)	$\alpha_1 = 3.2$	$\beta_1 = 0.8$
	$\alpha_2 = 2.6$	$\beta_2 = 1.3$
	$\alpha_3 = 1.7$	$\beta_3 = 2.8$
de Franceschi et al. (2009, unfiltered, along valley)	$\alpha_1 = 4.0$	$\beta_1 = 5$
	$\alpha_2 = 4.4$	$\beta_2 = 5$
	$\alpha_3 = 1.2$	$\beta_3 = 6$

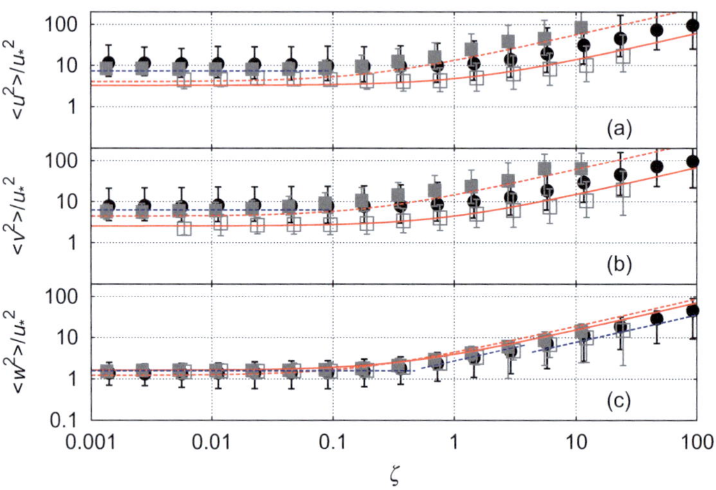

Fig. 3.9 $\langle u_i^2 \rangle / u_*^2$ as function of ζ. (**a**): $i = 1$; (**b**): $i = 2$; (**c**): $i = 3$. Observations from CABAUW (*full squares*: $z = 3$ m; *open squares*: $z = 60$ m) and from CCT (*full circles*: $z = 7.5$ m). *Red lines*: Eq. (3.58), with coefficients from de Franceschi et al. (2009) (*continuous*: along valley, filtered; *dashed*: along valley, unfiltered; see Table 3.3). *Blue lines*: stepwise behaviour from Kader and Yaglom (1990)

the original time series in order to separate the 'meteorological' unsteadiness from 'turbulence' affects the results by de Franceschi et al. (2009) as shown in Table 3.3. In Fig. 3.9 a few datasets and some parameterizations are reported. It results that the scaling is more effective on the vertical component than on the horizontal ones (which are more affected by the large scale parameters like h), and that the filtering reduces the variance.

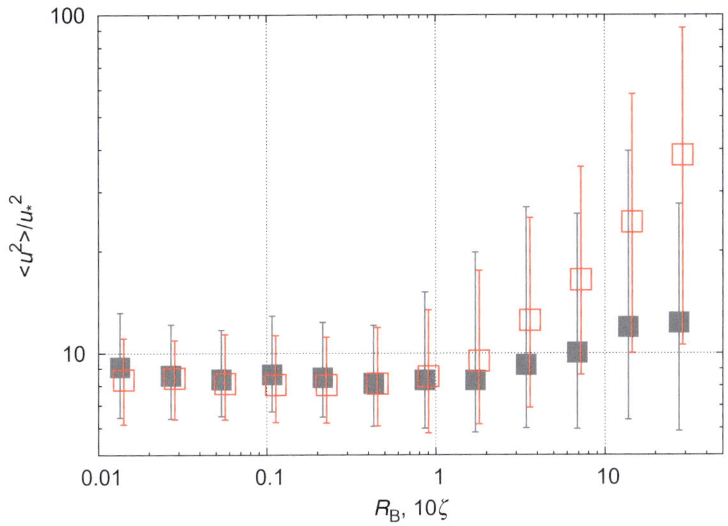

Fig. 3.10 $\langle u^2 \rangle / u_*^2$ as function of $R_{\rm B}$ (*full gray squares*) and of 10ζ (*open red squares*), from CABAUW data. The approximate, not universal, relation between $R_{\rm B}$ and ζ originates from Fig. 3.8

The self-correlation is expected to affect also these results, so it is worth to reexamine the problem using the gradient (or bulk) Richardson number. For a subset of the previous data, Fig. 3.10 shows the normalized variance of the streamwise component of the velocity as function of $R_{\rm B}(3, 60, 3, 60)$, following the definition (3.21). On the same plot, with a simple shift suggested by Fig. 3.8, the same data are reported as function of ζ: it is evident the increase at large instabilities in $[< u^2 > /u_*^2](\zeta)$, while $[< u^2 > /u_*^2](R_{\rm B})$ increases only slowly; the increase for $\zeta > 1$ is mostly due to self-correlation and the current parameterizations can be used with confidence up to moderate stability conditions.

3.4.3.2 The Third- and Fourth-Order Moments of the Velocity Components

It is more difficult to get reliable estimates of the odd moments than of the even ones; in general statistical moments of order larger than 2 are less well known than the variances.

Wind tunnel studies (Durst et al. 1987) show that in neutral conditions the probability density function of the normal to-the-wall component of velocity is not Gaussian, the asymmetry being related to the influence of the solid boundary (but not only). The thermal stratification also is expected to produce asymmetry: this point was investigated by Wyngaard and Coté (1971), by Chiba (1978) and discussed in Kader and Yaglom (1990), with special attention to the surface layer.

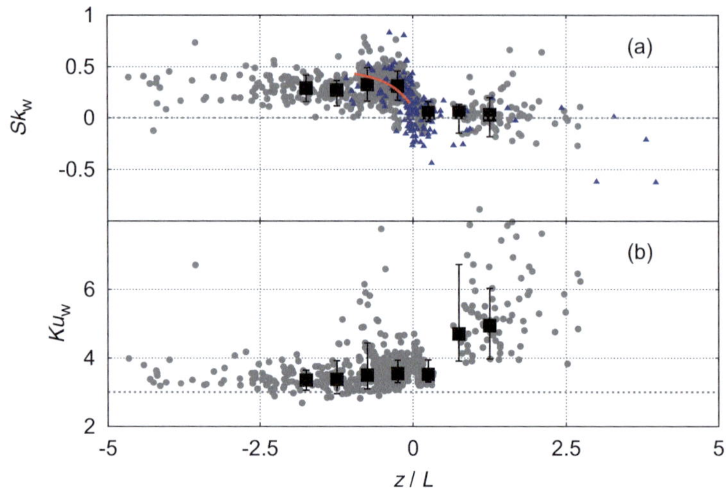

Fig. 3.11 (**a**): $Sk_w(z/L)$. *Gray dots*: observations; *black squares*: the same observations, binned at intervals of 0.5 z/L, from Barberis (2007). *Triangles*: observations redrawn from Chiba (1978). The *red line* corresponds to the empirical formula given by Chiba (1978, Eq. (7)). (**b**): $Ku_w(z/L)$. *Dots and squares* as in (**a**). The *horizontal dotted lines* indicate the values for a Gaussian distribution

Some results for heights outside the surface layer in unstable conditions will be presented in Sect. 3.6.1.

Long averaging times (necessary to minimize the statistical error and to account for the large, low frequency eddies typical of CBL) conflict with the requirement of steadiness of the time series. Data from the field experiment SGS 2002 (Higgins et al. 2007) elaborated by Barberis (2007) are reported in Fig. 3.11, together with some data from Chiba (1978), which show that the skewness of the vertical component of velocity is positive in unstable conditions, and almost zero in stable ones, with a sharp change at $z/L \approx 0$. The kurtosis is not far from the Gaussian value for unstable conditions, while becomes quite large (and scattered) for stable conditions.

The skewness of the horizontal components of the velocity shows a large scatter around zero, in atmospheric flows.

3.4.3.3 The Share of the TKE Among Components

The anisotropy induced by the presence of the surface and of the buoyancy is revealed by the ratios $\langle u_i^2 \rangle/(2E_k)$ and by their dependence on stability: the variance of the vertical velocity component increases as ζ increases, because of the increased intensity of the updraughts and downdroughts. The similarity functions (3.58) partially describe the data, as can be seen from Fig. 3.12.

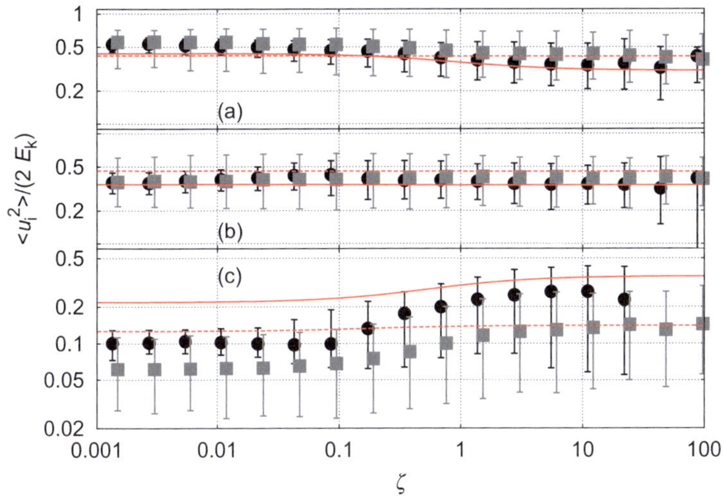

Fig. 3.12 Ratios of the wind component variances on the double of the TKE in the unstable boundary layer. (**a**): u component. (**b**): v component. (**c**): w component. *Full circles*: CABAUW data (at 3 and 60 m). *Full squares*: CCT data (at 7.5 m). *Lines* from relations (3.58) with the coefficients from de Franceschi et al. (2009, along-valley case), *continuous*: filtered; *dashed*: unfiltered

3.4.3.4 The Variance of the Temperature Fluctuations and the Horizontal Heat Flux

The variance of the temperature fluctuations scales on θ_*^2 and θ_{**}^2, in the quasi-neutral sublayer and in the intermediate and free-convective sublayers, respectively:

$$\frac{\langle \theta^2 \rangle}{\theta_*^2} = \alpha \ , \tag{3.59}$$

and

$$\frac{\langle \theta^2 \rangle}{\theta_*^2} = \beta \zeta^{-2/3} \ . \tag{3.60}$$

Some parameterizations are available in the literature (see, for instance, de Franceschi et al. 2009). A functional form which matches the scaling requirements is:

$$\frac{\langle \theta^2 \rangle}{\theta_*^2} = \frac{\alpha}{(1 + \beta \zeta)^{2/3}} \ , \tag{3.61}$$

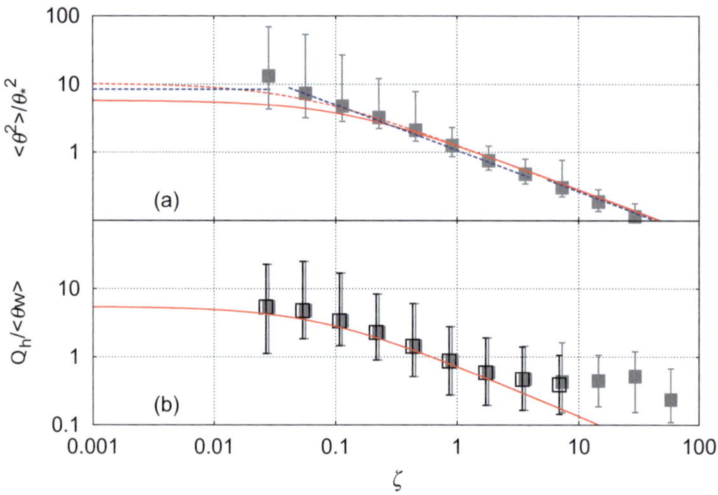

Fig. 3.13 (**a**) Variance of temperature fluctuations normalized over the temperature scale as function of ζ. *Gray symbols*: CCT observations, with a threshold value on the vertical heat flux of $0.04\,\mathrm{K\,m\,s^{-1}}$. *Red lines*: formula (3.61) with coefficients from de Franceschi et al. (2009): *continuous*: filtered; *dashed*: unfiltered. *Blue dashed lines*: from (3.59) and (3.60). Note in the intermediate sublayer the coefficient $\alpha_1 = 3.6$ while in the free-convective sublayer $\alpha_1 = 4.1$. (**b**) As in (**a**), but for the ratio of the horizontal to the vertical heat flux. *Gray symbols*: CCT observations, with a threshold value on the heat flux of $0.04\,\mathrm{K\,m\,s^{-1}}$; *black open symbols*: the same, with the additional filter of $u_* > 0.1\,\mathrm{m\,s^{-1}}$, as in Fig. 3.5. Equation (3.65) is reported by the *red line*, using mean velocity and temperature nondimensional gradients from Högström (1996)

and is reported in Fig. 3.13 together with some data, with the care to select cases with vertical heat flux over a threshold, to limit the influence of non-ideal effects. Coefficients according to different authors are reported in Table 3.4. As already noted, for conditions approaching neutrality (i.e. $\zeta \rightarrow 0$) $\theta_*^2 \rightarrow 0$ while the temperature fluctuations are still present, and the ratio $\langle\theta^2\rangle/\theta_*^2$ diverges for small ζ (Tampieri et al. 2009). Moreover, in the entire stability range, small vertical heat fluxes enhance the inhomogeneity effects.

As noted in Sect. 2.4.6, the existence of the horizontal heat flux in absence of any horizontal gradient of mean temperature is a relevant point in any approach based on the flux-gradient relationship, such as most of the closures. Wyngaard et al. (1971) and Monin and Yaglom (1971) discuss this point.

According to Kader and Yaglom (1990), in the dynamic sublayer, the horizontal heat flux scales on u_* and θ_*, in the intermediate sublayer on u_{**} and θ_{**}, and in the free-convective sublayer on w_* and θ_{**}. In other words, the horizontal heat flux is proportional to the vertical heat flux in the dynamic and free-convective sublayers, with different proportionality coefficients. It results:

$$\frac{Q_\mathrm{h}}{\langle\theta w\rangle} = \alpha_1 \;, \tag{3.62}$$

Table 3.4 Numerical coefficients for the variance of the temperature and for the horizontal heat flux (unstable conditions)

$\langle\theta^2\rangle$			
Kader and Yaglom (1990)	$\alpha = 8.4$	$\beta = 4.1$	(for $\zeta < 2\,\beta = 3.6$)
de Franceschi et al. (2009, filtered, along valley)	$\alpha = 6.0$	$\beta = 9.2$	
de Franceschi et al. (2009, unfiltered, along valley)	$\alpha = 10.4$	$\beta = 22.5$	
$\langle\theta u\rangle$, Q_h			
Wyngaard et al. (1971) for $\langle\theta u\rangle$	$\alpha_4 = 5$		
Kader and Yaglom (1990) for $\langle\theta u\rangle$	$\alpha_1 = 3.8$		
	$\alpha_2 = 0.7$		
	$\alpha_3 = 0.4$[a]		
Equation (3.65) for Q_h, Högström (1996)	$\alpha_4 = 5.7$		
Equation (3.65) for Q_h, Kader and Yaglom (1990)	$\alpha_4 = 4.9$		

[a]Estimated by eye, from their Fig. 3

$$\frac{Q_h}{\langle\theta w\rangle} = \alpha_2 \zeta^{-2/3} , \tag{3.63}$$

and

$$\frac{Q_h}{\langle\theta w\rangle} = \alpha_3 , \tag{3.64}$$

where $Q_h = \left(\langle\theta u\rangle^2 + \langle\theta v\rangle^2\right)^{1/2}$ is the horizontal heat flux. Note that the above-cited authors refer to the along-wind component $\langle\theta u\rangle$ of the horizontal heat flux, assuming that the transversal component is zero.

According to (2.86), the horizontal heat flux occurs in presence of vertical gradients of mean velocity and temperature, thus Wyngaard et al. (1971) suggest:

$$\frac{Q_h}{\langle\theta w\rangle} = \alpha_4 \Phi_U \phi_\Theta , \tag{3.65}$$

which can be applied using the non-dimensional gradient formulas discussed above: see Fig. 3.13b and the coefficients as reported in Table 3.4. Note that the correct behaviour for free-convective conditions cannot be obtained from (3.65) which is essentially based on the flux-gradient relationship, while the behaviour of the free-convection sublayer is non-diffusive. In this sublayer, only the scale reasoning holds, so the correct behaviour is given by (3.64).

3.4.3.5 The Temperature Skewness

There are few data about the third-order moment of the temperature, which is also an indication for passive tracers. The results are quite scattered: see Fig. 3.14. Kader and Yaglom (1990, Figs. 3 and 5) suggest S_k to be about 0.4 in near neutral

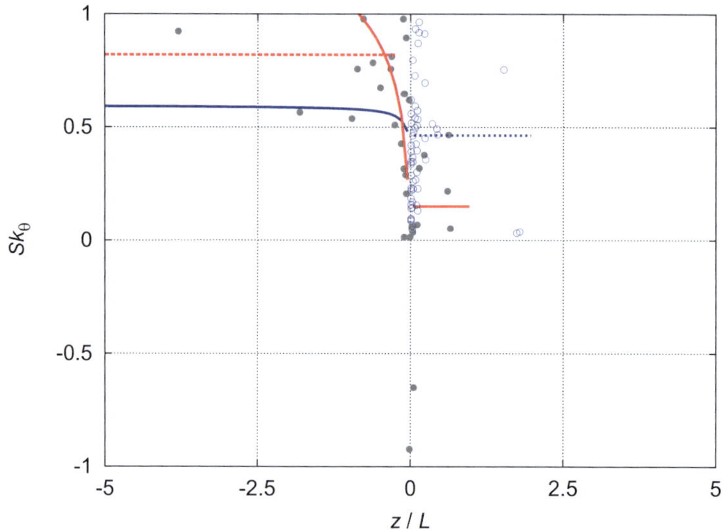

Fig. 3.14 Temperature fluctuation skewness $Sk_\theta(z/L)$. *Gray dots*: observations redrawn from Andreas et al. (1998, Fig. 11). *Black open circles*: observations from Dias et al. (1995). The *red line* corresponds to the empirical formulas suggested by Andreas et al. (1998, Eqs. (12) and (13)). The *blue continuous line* is derived interpolating from Kader and Yaglom (1990); the blue dotted line refers to Dias et al. (1995)

conditions and 0.7 as instability increases. Andreas et al. (1998, Fig. 11) suggest that in stable conditions the skewness is about 0 while in unstable condition it increases up to values near 1 (obviously constant with z/L) in free convective conditions. Measurements by Dias et al. (1995) are also reported for the stable case, being the average value about 0.4. It is evident that our knowledge about the skewness is quite poor, because of the difficulty to reach steady conditions and thus to make long time averages.

3.4.3.6 The Dissipation of the TKE

From (3.10), in the surface layer approximation, neglecting the divergence of the third-order moments, it results:

$$\frac{\kappa \varepsilon z}{u_*^3} = \Phi_{\mathrm{U}}(\zeta) + \zeta \,, \tag{3.66}$$

that is, for quasi-neutral conditions $\varepsilon \propto u_*^3/z$ and for convective conditions $\varepsilon \propto u_*^3/|L|$.

Similarly, from (2.75), it results:

$$\frac{\kappa \varepsilon_\Theta z}{u_* \theta_*^2} = \Phi_\Theta(\zeta) \,, \tag{3.67}$$

giving $\varepsilon_\Theta \propto \zeta^{-4/3}$. This result was already reported by Wyngaard and Coté (1971) and by Högström (1990).

The proportionality constants are evaluated from the observations, and are often different from those that could be obtained from the nondimensional gradients of the mean quantities. A variety of values can be found in the literature: see for instance Albertson et al. (1997), Frenzen and Vogel (2001), Charuchittipan and Wilson (2009).

3.5 The Outer Region in Neutral Conditions

A few considerations about the structure of the PBL above the surface layer are developed here.

3.5.1 *The Mean Velocity in the Ekman Layer*

To derive the profile of the mean velocity in neutral conditions, the simplest approach is to use (3.14), neglect the viscous flux of momentum ($\nu = 0$), and assume the flux-gradient relationship (2.80) for each horizontal direction with a constant eddy diffusion coefficient (in complex variables, $\mathbf{X} = -K_\tau \,\mathrm{d}\mathbf{V}/\,\mathrm{d}z$). The boundary conditions are:

$$\mathbf{V}(0) = 0 \ , \tag{3.68}$$

$$\lim_{z\to\infty} \mathbf{V}(z) = \mathbf{V}_\mathrm{g} \ . \tag{3.69}$$

For the solution, see the Exercise 3.4.

The mean velocity profile takes the form:

$$\mathbf{V} = \mathbf{V}_\mathrm{g} \left\{1 - \exp\left[-(1+\imath)(f/2K_\tau)^{1/2} z\right]\right\} \ . \tag{3.70}$$

Orienting the coordinate axes so that $U_\mathrm{g} = |\mathbf{V}_\mathrm{g}|/\sqrt{2}$ and $V_\mathrm{g} = -|\mathbf{V}_\mathrm{g}|/\sqrt{2}$, it results:

$$\frac{\mathrm{d}U}{\mathrm{d}z}|_{z=0} = |\mathbf{V}_\mathrm{g}|(f/K_\tau)^{1/2} \ , \quad \frac{\mathrm{d}V}{\mathrm{d}z}|_{z=0} = 0 \tag{3.71}$$

i.e. the mean velocity near the surface is rotated anticlockwise of $\pi/4$ with respect to the direction of the geostrophic wind, in the Northern hemisphere.

From (3.70) we can estimate that the mean velocity approximates the geostrophic wind at an height $\beta(f/2K_\tau)^{-1/2}$. From (3.16) the neutral boundary layer depth has been estimated to be $\alpha u_*/f$, so that the eddy diffusion coefficient in the Ekman

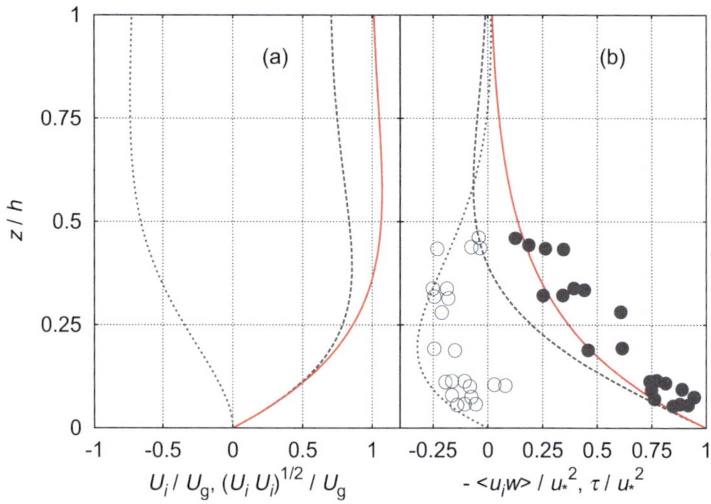

Fig. 3.15 (**a**) Vertical profiles of the horizontal velocity components in direction x (*dashed line*), in direction y (*dotted line*) and of the modulus of the velocity (*continuous red line*) normalized over the modulus of the geostrophic velocity (in this example, $V_g = -U_g$). (**b**): Vertical profiles of the negative of the components of the vertical momentum flux (*dashed line*: x component; *dotted line*: y component) and of the modulus of the momentum flux (*continuous red line*) normalized over its surface value u_*^2. The *full and open symbols* are observations of the x and y components of the momentum flux over the sea, redrawn from Garratt (1992). Here $\alpha = 0.3$ and $\beta = 4$

model turns out to be $K_\tau = \alpha^2/(2\beta^2)\, u_*^2/f$ and the ratio between the geostrophic wind velocity and the friction velocity is $|\mathbf{V}_g|/u_* = \sqrt{2}\beta/\alpha$.

This solution is represented in Fig. 3.15 with $\alpha = 0.3$ and $\beta = 4$. Some data of vertical momentum flux are also reported.

3.5.2 *Truly-Neutral and Conventionally-Neutral Boundary Layers*

The formula (3.16) accounts for the combined effects of the friction velocity (that enhances mixing and thus increases the boundary layer depth) and of the rotation (that limits it). This formula is consistent with a diffusion-like approach, in which the diffusion coefficient is K_τ and the time scale is f^{-1}:

$$h^2 \propto \frac{K_\tau}{f} , \tag{3.72}$$

and the eddy diffusion coefficient is given by the product of a velocity scale and a mixing length scale:

$$K_\tau \propto u_* h . \tag{3.73}$$

As widely discussed (Zilitinkevich and Esau 2002; Zilitinkevich et al. 2007; Mironov and Fedorovich 2010), the neutral boundary layer depth can be limited also by the presence of a stably stratified troposphere above it: this is the case of the 'conventionally neutral' boundary layer, different from the 'truly neutral' boundary layer, in which the whole atmosphere is supposed to be neutrally stratified, and (3.16) holds. Note that the most common case in the atmosphere is the conventionally neutral one (the troposphere is stably stratified). To take into account the stability above, the mixing length scale can be written as u_*/N, the eddy diffusion coefficient turns out to be:

$$K_\tau \propto \frac{u_*^2}{N}\,, \tag{3.74}$$

and the consistent estimate of the NBL depth is:

$$h \propto \frac{u_*}{\sqrt{Nf}}\,. \tag{3.75}$$

The NBL depth can be estimated by the interpolation formula:

$$h^{-2} = \frac{f^2}{C_1^2 u_*^2} + \frac{Nf}{C_2^2 u_*^2}\,, \tag{3.76}$$

where the coefficients are determined from the observations (Zilitinkevich et al. 2007; Zilitinkevich and Esau 2007): $C_1 \approx 0.6$, $C_2 \approx 1.36$.

The smallest term dominates the resulting depth. The effect of the stratification can be quite relevant: for a range of N from 0 to $0.01\,\mathrm{s}^{-1}$ characteristic of atmospheric flows, and for $f = 10^{-4}\,\mathrm{s}^{-1}$, the ratio of the effective depth h over the 'truly neutral' estimate decreases from 1 to 0.2, suggesting the possibility of neutral boundary layers shallower than the estimate (3.16).

3.5.3 *Resistance Laws*

The PBL resistance laws relates the surface fluxes to the free-troposphere dynamics. Here we limit to neutral, barotropic boundary layers. The geostrophic drag is defined [similarly to (3.32)]:

$$c_g = \frac{u_*}{|\boldsymbol{U}_g|}\,, \tag{3.77}$$

and the resistance laws become (Zilitinkevich and Esau 2005):

$$\frac{\kappa}{c_g}\cos\alpha_0 = \ln\frac{h}{z_0} - A\,, \tag{3.78}$$

$$\frac{\kappa}{c_g} \sin \alpha_0 = -\frac{fh}{u_*} B , \tag{3.79}$$

where α_0 is the cross-isobaric angle (the angle between the surface shear stress and the geostrophic wind ($\alpha_0 = \pi/4$ for the Ekman solution).

The discussion of the dependence of the coefficients A and B on the boundary layer depth and on the stability conditions goes beyond the scope of the book. However, in the case of neutral conditions, an hint can be obtained using the results of the previous section. For truly-neutral boundary layers, the depth is given by (3.16) and the parameter A reads:

$$A = -\frac{\kappa}{c_g} \cos \alpha_0 + \ln(c_g R_{o0}) + \ln \alpha , \tag{3.80}$$

while if the tropospheric stability dominates, using (3.75) with the coefficient C_2 as in (3.76), it results:

$$A = -\frac{\kappa}{c_g} \cos \alpha_0 + \ln \left(\frac{f^{1/2}}{N^{1/2}} c_g R_{o0} \right) + \ln C_2 , \tag{3.81}$$

where the Rossby number based on the roughness length appears:

$$R_{o0} = \frac{|\boldsymbol{U}_g|}{f z_0} . \tag{3.82}$$

The dependence of A (and of B too) on the characteristics of the PBL was outlined by Zilitinkevich and Esau (2002), referring to the variability of the classical estimates of these parameters between atmospheric measurements (which typically refer to conventionally neutral boundary layers) and laboratory and numerical simulations (for truly-neutral boundary layers).

3.6 Some Features of the Convective Boundary Layer

Above the surface layer, the convection is expected to dominate the structure of the convective boundary layer. The large vortices, qualitatively shown by the SODAR and LIDAR measurements (see Figs. 1.2, 1.3 and 1.4 in the Introduction) and widely investigated by different observational and numerical techniques (see, for instance, Lenschow et al. 2012), are characterized by a vertical length scale of the order of the boundary layer depth h. From (3.35), the velocity scale becomes:

$$W_* = w_*(h) , \tag{3.83}$$

where the reference temperature θ_{00} is substituted by the temperature averaged over the layer depth:

$$\overline{\Theta} = h^{-1} \int_0^h \Theta_f \, dz \qquad (3.84)$$

(Deardorff 1970; Hunt et al. 1988). Because of the intense mixing, the mean velocity and temperature are approximately constant with height, above the surface layer (see, for instance, Ansmann et al. 2010). Some temperature profiles are reported in Fig. 3.16, which refers to the same case of Fig. 1.4. In Fig. 3.17 the sensible and latent turbulent heat fluxes near the surface for the same day are also reported, together with the inverse of the Obukhov length. The diurnal cycle is evident, as well as the dominance of buoyancy effects over the dynamic ones during daytime. Further profiles of the mean temperature can be found in the literature: see, for instance, Angevine et al. (1998). Near the top (for $z \approx h$), the temperature increases and mixing between the boundary layer and the troposphere can occur (entrainment region).

The CBL is affected by the presence or absence of moist convection leading to clouds at the top (see, for instance Ansmann et al. 2010; Lenschow et al. 2012). Some features of the entrainment region have been only recently investigated (thanks to the improvements in the available instrumentations): see Malinowski et al. (2011) for a pair of case studies.

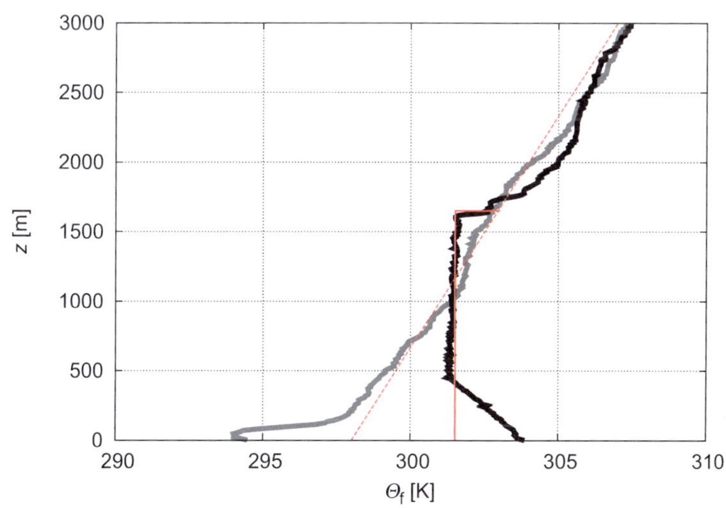

Fig. 3.16 Temperature profiles measured at the same site and time as in Fig. 1.4 (courtesy G. Bonafe, ARPA ER, Bologna, Italy) for 06 h (*gray*) and 12 h (*black*) local time. The *red lines* represent the idealized situation as treated in the slab model, Sect. 3.6.2

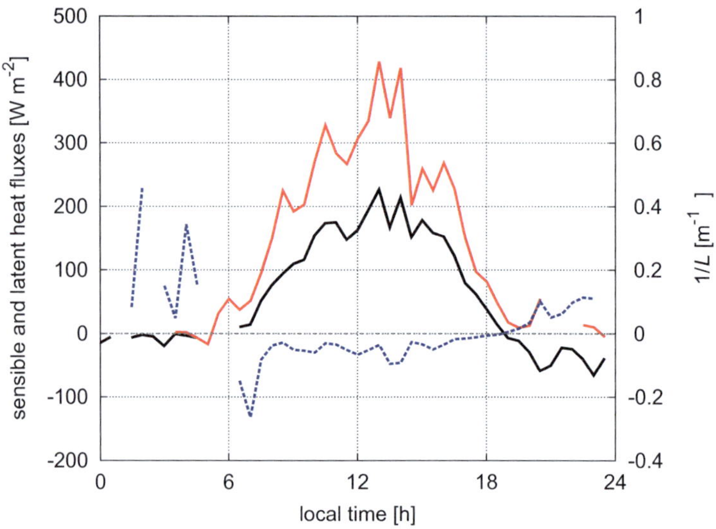

Fig. 3.17 Sensible (*black line*) and latent (*red*) turbulent heat fluxes, and inverse of the Obukhov length (*blue dashed line*) measured at the same site and time as in Fig. 1.4 (courtesy G. Bonafe, ARPA ER, Bologna, Italy)

3.6.1 Second- and Third-Order Moments of Fluctuations

The large convective eddies are characterized by updrafts and downdrafts (the updrafts being of smaller horizontal section and of larger vertical velocity than the downdrafts), resulting in the variance of the vertical velocity with a maximum in the cental part of the boundary layer, in the positive skewness (the probability density function of the vertical velocity has a positive tail) increasing from the surface value of 0.4 to ≈ 0.8, and in the kurtosis increasing from 3 to 4.

Some formulas describing the variance of the velocity components and of the temperature can be given, appropriate for the overall features of the CBL (and thus neglecting the specific surface-layer behaviour discussed in Sect. 3.4.3). For the streamwise component of the velocity, Strunin et al. (2004) suggest:

$$\frac{\langle u^2 \rangle}{W_*^2} = 1.4\xi^{4/3}\left(1 - 0.7\xi\right)^{2/3} + 0.05\xi^{-2/3}\left(1 - 0.8\xi\right)^{2/3} , \tag{3.85}$$

where $\xi = z/h$ (see also Sorbjan 1991). For the vertical component, Lenschow et al. (1980) give:

$$\frac{\langle w^2 \rangle}{W_*^2} = 1.8\xi^{2/3}(1 - 0.8\xi)^2 . \tag{3.86}$$

A formula for the variance of the temperature fluctuations has been suggested by Strunin et al. (2004):

$$\frac{\langle \theta^2 \rangle}{\Theta_{**}^2} = 1.8\xi^{-2/3}\left(1-\xi\right)^{4/3} + 1.4\xi^{4/3}\left(1-\xi\right)^{-2/3} , \tag{3.87}$$

where

$$\Theta_{**} = \theta_{**}(h) , \tag{3.88}$$

with the same choice as before for the reference temperature (Deardorff 1970).

In the approximation of mean temperature almost constant with height and increasing in time, Eq. (3.3), where viscosity is neglected, suggests that the heat flux decreases linearly with the height: $\langle \theta w \rangle(z) = \langle \theta w \rangle_0 (1 - \alpha\xi)$, fairly consistent with the data, Fig. 3.20b.

Lenschow et al. (1980) give a formula for the third-order moment of the vertical component of velocity:

$$\frac{\langle w^3 \rangle}{W_*^3} = 1.2\xi(1 - 0.7\xi)^3 , \tag{3.89}$$

that together with (3.86) gives the vertical profile of the skewness.

ARTIST-CBL data (see Gryanik and Hartmann 2002, Table 1) are reported in Figs. 3.18, 3.19 and 3.20 together with the relationships presented above. The fourth-order moment is quite variable for the horizontal components (not shown), while for the vertical component it shows a maximum in the centre of the CBL; the fit reads:

$$\frac{\langle w^4 \rangle}{W_*^4} = 8.4\xi^{4/3}(1 - 0.6\xi)^4 . \tag{3.90}$$

3.6.2 *The Morning Growth of the CBL*

At mid latitudes the time evolution of the PBL is driven by the diurnal cycle.

In conditions of intense convection the flux Richardson number is large (in absolute value) and the mean wind shear is small: it is a case of the shear-free boundary layers (see, for instance, Zilitinkevich et al. 1998). Under such conditions, the main features of the morning growth of the boundary layer can be described by a simple approach.

The heat flux decreases with height, consistently with the temperature change (in time). From (3.3), neglecting molecular viscosity, integrating over the boundary layer depth, and neglecting the difference between the mean temperature at the top

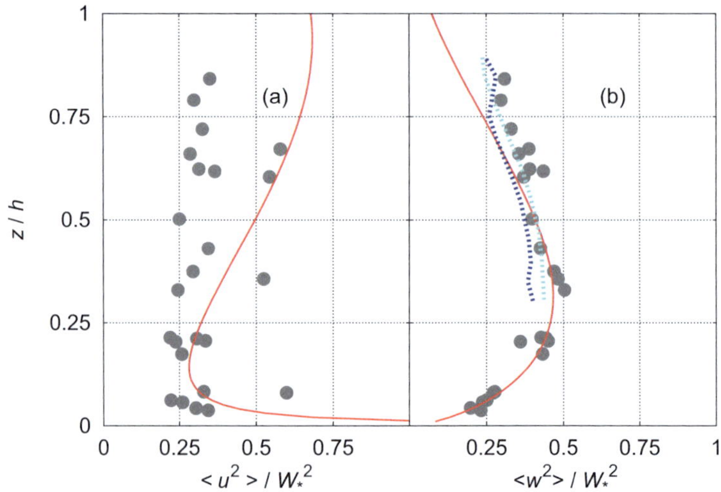

Fig. 3.18 (**a**) Vertical profiles of the variance of the streamwise velocity component normalized over W_*^2. *Dots*: ARTIST-CBL data. *Red line*: Eq. (3.85). (**b**) The same as in (**a**), but for the variance of the vertical velocity component normalized over W_*^2. *Dots*: ARTIST-CBL data; *dashed lines*: data from the LIFT experiment, redrawn from Lenschow et al. (2012): *blue line*, weak convection, *cyan line*: strong convection. *Red line*: Eq. (3.86)

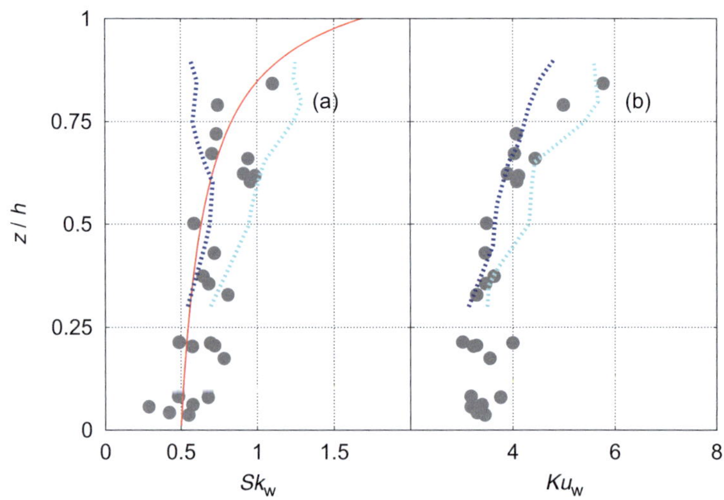

Fig. 3.19 (**a**) Vertical profiles of the skewness of the vertical velocity component. *Dots*: ARTIST-CBL data; *dashed lines*: data from the LIFT experiment, redrawn from Lenschow et al. (2012): *blue line*, weak convection; *cyan line*, strong convection. (**b**) The same as in (**a**), but for the kurtosis of the vertical velocity component. The *red line* in (**a**) is derived from (3.89) and (3.86)

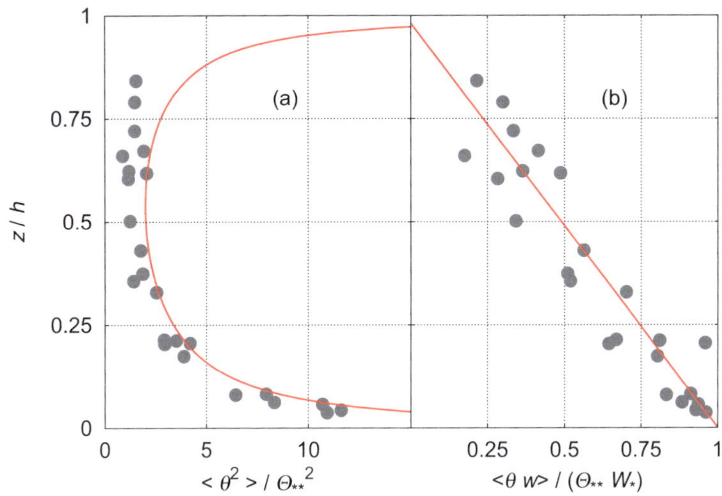

Fig. 3.20 (**a**) Vertical profiles of the temperature variance normalized over Θ_{**}^2. *Dots*: ARTIST-CBL data; *continuous red line*: from (3.87). (**b**) Vertical profile of the heat flux normalized over its surface value. *Dots*: ARTIST-CBL data; *continuous red line*: best fit of the data, showing that the heat flux goes to zero at the CBL top

$\Theta_f(h)$ and the temperature averaged over the layer $\overline{\Theta}$, it results:

$$h\frac{d\overline{\Theta}}{dt} = \langle\theta w\rangle_0 - \langle\theta w\rangle(h) \ , \tag{3.91}$$

which states that the time changes of the averaged temperature depend on the balance of the heat flux in the layer itself. The time scale of these changes is h/W_*.

Let consider the idealized situation, depicted in Fig. 3.16, characterized by a linear profile of the background temperature: its slope is $\Gamma = d\theta_0/dz$. Define the temperature jump at the top of the layer $\Delta\Theta = \theta_0(h) - \overline{\Theta}$. If $\Delta\Theta \neq 0$ a downwards (negative) heat flux occurs at the boundary layer top, which in analogy with the flux-gradient relationship is written in term of an entrainment velocity w_e:

$$\langle\theta w\rangle(h) = -w_e \Delta\Theta \ , \tag{3.92}$$

In the absence of meteorological forcing (for instance, the subsidence), the entrainment velocity corresponds to the rate of change of the boundary layer depth:

$$\frac{dh}{dt} = w_e \ , \tag{3.93}$$

and substituting in (3.91) with (3.92) the evolution equation for the temperature jump results:

$$\frac{\mathrm{d}\Delta\Theta}{\mathrm{d}t} = \Gamma w_\mathrm{e} - \frac{\langle\theta w\rangle_0}{h} - \frac{\Delta\Theta}{h} w_\mathrm{e} \,. \tag{3.94}$$

A parameterization of the turbulent exchange at the top, i.e. an empirical formulation for the entrainment velocity, is necessary to solve the system made by (3.93) and (3.94). This system gives us the so-called slab model.

Various formulations for w_e have been proposed in the literature. Here we consider two possibilities. Note that LES are widely used for simulations of the CBL, as we shall discuss in Sect. 6.4.

3.6.2.1 The 'Encroachment' Model

The simplest model is obtained by eliminating w_e from (3.93) using (3.94) and assuming $\Delta\Theta = 0$ (but $\mathrm{d}h/\mathrm{d}t > 0$). It results:

$$\frac{\mathrm{d}h}{\mathrm{d}t} = \frac{\langle\theta w\rangle_0}{\Gamma h} \,, \tag{3.95}$$

and

$$h(t) = \sqrt{\frac{2}{\Gamma}\int_0^t \langle\theta w\rangle_0 \,\mathrm{d}t'} \,, \tag{3.96}$$

i.e. the time evolution of the depth is determined only by the surface heat flux and by the background temperature gradient.

3.6.2.2 The Effects of Surface Friction and of the Entrainment at the Top

More complex models are necessary to deal with the effects of the surface friction, which may be relevant for the initial development of the boundary layer (for instance, in the early morning, if the diurnal cycle occurs), and with the effect of entrainment of warm air from above. A simple representation of these effects is given, for instance, by this formulation for the flux at $z = h$:

$$\langle\theta w\rangle(h) = -A\langle\theta w\rangle_0 - B\frac{u_*^3\theta_{00}}{gh} = -\langle\theta w\rangle_0\left(A - B\kappa\frac{L}{h}\right) \,, \tag{3.97}$$

introduced by Batchvarova and Gryning (1990) as a simplification of a Zilitinkevich (1975) formula. Suggested values are $A = 0.2$ and $B = 2.5$. Note that this formula is equivalent to the expression used by Luhar (1998).

Let us consider the two extreme cases, describing respectively the mature evolution of the CBL and its initial development, as friction rather then convection dominates.

For pure convection, $B = 0$, and we obtain an equation like (3.95), with a coefficient $(2A + 1)$ that multiplies the surface heat flux: the growth of the CBL height is enhanced by the heat flux at the top ($A > 0$).

For neutral conditions, $A = 0$, and a slower growth results:

$$\frac{\mathrm{d}h}{\mathrm{d}t} = \left[\frac{2\, B u_*^3 \vartheta_{00}}{g \Gamma h^2} \right] . \tag{3.98}$$

Note that in the limit of large Obukhov length (surface heat flux small but non zero), the same solution (3.98) results.

Furthermore, it must be observed that the entrainment coefficient A is function of the dynamics of the boundary layer (Zilitinkevich 2012, Sect. 13.3.4), so that more complex models are suited for applications where the CBL depth is critical.

3.6.3 The Day-Night Transition and the Residual Layer (RL)

When the forcing due to the surface heat flux begins to decrease (say in the afternoon), some important changes occur in the CBL: in particular, the TKE decays, but the convective vertical up- and down-drafts continue to mix the boundary layer. At sunset the forcing stops, and the decay occurs faster.

Early observations are reported in Caughey and Kaimal (1977); a refined, more recent description is based on field campaigns, like TRANSFLEX (Fernando et al. 2013) and BLLAST (Lothon et al. 2014), designed specifically for this purpose, and generically on extensive measurements of mid-latitude PBL evolution, like CASES99 (Poulos et al. 2002) or LITFASS-2003 (Beyrich and Mengelkamp 2006). See also a comparison between two experiments in Sastre et al. (2015), which gives an idea of the observation variability, which does not help for a general understanding of the phenomenology. Numerical simulations by Nieuwstadt and Brost (1986), Sorbjan (1997), Darbieu et al. (2015) (among others) integrate the picture.

Observations of mean temperature profiles show that the almost constant profile of the previous hours is maintained also after sunset, above the SBL which develops at the ground, as shown by Poulos et al. (2002, Fig. 18) or Lothon et al. (2014, Fig. 11): this corresponds to a well-mixed layer without forcing from below, called the residual layer (RL).

The decay of the turbulence, when the forcing produced by the surface heat flux decreases in the afternoon and possibly becomes negative, is documented by the observations in Nadeau et al. (2011, Figs. 7, 9, 12) or by Lothon et al. (2014, Fig. 12). It is worth noting that a fast decrease of the forcing occurs during solar eclipses, and the effect on TKE is documented by Anfossi et al. (2004) or by Bala Subrahamanyam et al. (2011).

A further aspect is the length scale of the eddies in the RL: the decrease of the surface forcing tends to maintain the larger scales while the smaller ones dissipate quickly: this point has been re-examined recently by Darbieu et al. (2015), but it was pointed out by the LES simulations by Sorbjan (1997). Analogous results (the scale of the most energetics eddies does not change) are obtained by Goulart et al. (2003) using a spectral model.

The decay of TKE in a convectively-driven boundary layer is described by (3.10), simplified by retaining the heat flux production and dissipation and neglecting the viscous diffusion, the shear production and the third-order terms divergence (Nadeau et al. 2011).

For the depth-averaged TKE $\overline{E_k}$, the evolution equation reads:

$$\frac{d\overline{E_k}}{dt} = A\frac{g\langle w\theta\rangle_0(t)}{\theta_{00}} - \varepsilon = A\frac{g\langle w\theta\rangle_0(t)}{\theta_{00}} - C_\varepsilon\frac{\overline{E_k}^{3/2}}{h}\,, \tag{3.99}$$

where A is an empirical constant and the dissipation is parameterized in terms of TKE and CBL depth, following a suggestion by Nieuwstadt and Brost (1986), with $C_\varepsilon = 2$.

Putting $A = 0$, (3.99) describes the time evolution of TKE, if the forcing is switched off at $t = 0$ (sharp switch-off). Assuming h constant and $E_k(0) = CW_*^2$ it results:

$$\frac{\overline{E_k}}{W_*^2} = \left(\frac{C_\varepsilon}{2}\frac{W_* t}{h} + \frac{1}{C^{1/2}}\right)^{-2}\,. \tag{3.100}$$

The sharp switch-off of the heat flux at the ground is not realistic: thus it is important to properly represent the progressive decrease. This decay can be represented with various functions; here a cosine-squared function is used:

$$\frac{\langle w\theta\rangle_0(t)}{\langle w\theta\rangle_0(0)} = \cos^2\left(\frac{\pi t}{2T}\right)\,, \tag{3.101}$$

where T represents the decay time of the heat flux from its maximum value at $t = 0$, that at mid-latitudes corresponds to midday. The TKE decay estimated by the numerical integration of (3.99) is compared with some experimental observations in Fig. 3.21.

3.7 Stable Boundary Layers

As outlined in the discussion about the TKE evolution equation (3.10), increasing stability reduces the TKE and the mixing, in broad sense, leading to guess the existence of the critical flux Richardson number as an upper limit, beyond which

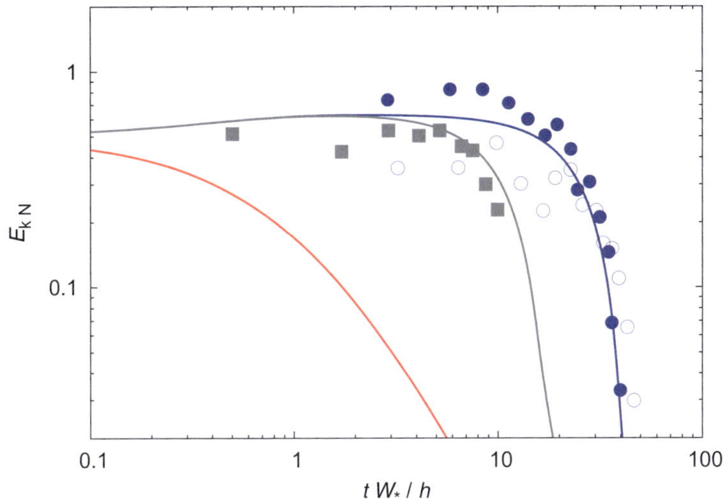

Fig. 3.21 Normalized averaged TKE $E_{kN} = \overline{E_k}/W_*^2$ as function of the nondimensional time, for the day-night transition. *Blue symbols*: data redrawn from Nadeau et al. (2011, Figs. 7a: open and 7c: full); *gray symbols*: data from Anfossi et al. (2004, Fig. 3) (note that these data have been rescaled with respect to the original figure). *Red line*: Eq. (3.100); *blue line*: Eq. (3.99) with $A = 1$ and a cosine-square modulation of the surface heat flux, with decay time in (3.101) of about 7.5 h, proper for Nadeau et al. (2011) observations; *gray line*: the same as before, with $T \approx 1.3$ h, for Anfossi et al. (2004)

turbulence modifies its features or even disappears. Reducing the mixing means that the vertical fluxes of momentum and heat become smaller as stability increases: the exchanges between surface and atmosphere reduce, possibly becoming almost negligible. As a consequence, the velocity and temperature scales u_* and θ_*, as well as the Obukhov length, reduce their relevance (and reliability) as stability increases. Last, but not least, the concept of boundary layer as a layer whose dynamics is driven by the exchanges between surface and atmosphere weakens: effects related to processes occurring aloft are expected to become important. The depth of the SBL is related to a number of factors; the part of the SBL affected by surface fluxes may be very shallow, and in such circumstances the surface layer almost disappears.

To cope with the complexities of the SBL, it is useful to take into account the following aspects:

- a distinction between moderate and strong stability can be made (Mahrt 2014).

  In the moderately stable boundary layers the turbulent fluxes of momentum and heat are relevant, so that the related scales of velocity, temperature and length can be used, according to MOST. To face with the changes of the fluxes with height, the local similarity approach has been forwarded (Nieuwstadt 1984): the scales must be considered as local quantities, so the velocity scale is $\tau(z)^{1/2}$, the temperature scale is $\langle\theta w\rangle(z)/\tau(z)^{1/2}$, and the local value of the Obukhov length is

$$\Lambda(z) = -\frac{\Theta_r\, \tau(z)^{3/2}}{\kappa\, g\, \langle\theta w\rangle(z)}\,, \tag{3.102}$$

where Θ_r is for instance the mean temperature at height z. Note that the local similarity relates to the z-less concept, i.e. the independence of similarity laws from the height (Wyngaard and Coté 1972).

Local stability can be also measured in terms of the gradient Richardson number (3.20): being based on the gradients of the mean quantities (that can be large in conditions of reduced mixing) it is expected to be a measure of stability better than Λ in strongly stable boundary layers. The existence of a critical value of the flux Richardson number (Sect. 3.1) suggests that the same should occur for the gradient Richardson number;

- based on the height dependence of turbulence parameters, a distinction between traditional and upside-down SBL can be made (Mahrt and Vickers 2002; Banta et al. 2006; Tampieri et al. 2015).

  We can define the traditional SBL (tSBL) as those SBL with τ decreasing and $\langle\theta w\rangle$ increasing with height (considered with its sign; note that in absolute value the heat flux decreases with height): broadly speaking, the tSBL can be said to be affected mainly by the surface fluxes. For tSBL, the depth can be determined, taking the height at which the fluxes reduce of a given fraction of their surface values (this issue will be dealt with in Sect. 3.7.4).

  On the other hand, as the surface fluxes are negligible or not relevant for the dynamics, while the TKE is mainly produced aloft (for instance, by the wind shear), we can define the upside-down SBL (uSBL) as those SBL with both τ and E_k increasing with height. For these SBL, the depth can be related to some features of the mean profiles, for instance to the occurrence of a low-level jet;
- besides the TKE, the TPE has to be accounted for (Zilitinkevich et al. 2008) to deal with the evolution of the SBL.

  In detail, as the stability increases, the TPE increases while the TKE decreases: thus the TPE represents a reservoir of energy, which is made available to the system as the stability decreases. In a unsteady SBL, this interplay allows to maintain the turbulence also during periods of high Richardson number (higher than some critical value). Moreover, the stability affects the partition between horizontal and vertical components of the velocity, so that the ratios $\langle u_i^2\rangle/(2E_k)$ with $i = 1, 2, 3$, are function of stability, with a behaviour different from that of unstable conditions;
- steadiness is almost impossible in strongly stable conditions. Decaying or intermittent turbulence is a common feature;
- in a stable environment wavelike motions are a common presence. Gravity waves are the prototype of such phenomena; however the disturbances are often too short in time to speak safely of waves (and, showing that we need to refer

to simple model, in the literature the term ‘dirty wave’ is used to tell about oscillation which cannot be considered periodic). Wavelike motions are difficult to observe because often they are superimposed to mechanical fluctuations (mechanical turbulence) in spectral terms. Internal gravity waves transport vertically momentum but not heat; in general they produce an increase of (turbulent) potential energy, which is possibly transferred to TKE. The non-local nature of the wavelike motions emphasizes the difficulty to deal with turbulence in the SBL in terms of local variables; in particular, this observation explains why the Richardson number or the Obukhov length cannot fully characterize turbulence in stable conditions, and thus why the similarity approach fails, when vertical fluxes are too small to dominate turbulence. Besides the theoretical interest, this issue is of overwhelming importance for numerical simulations of weather, climate and atmospheric composition (Sun et al. 2015).

A picture of the vertical profiles of the second-order moments used to define tSBL and uSBL is reported in Figs. 3.22 and 3.23, where the momentum and heat fluxes and the TKE are made nondimensional using the respective surface value, and the height is made nondimensional using as scale height (which is also the empirical estimation of the SBL depth for tSBL) the average of the vertical variation scale between momentum flux and heat flux for tSBL, and the average between momentum flux and TKE for uSBL. (The variation scale can be defined in different

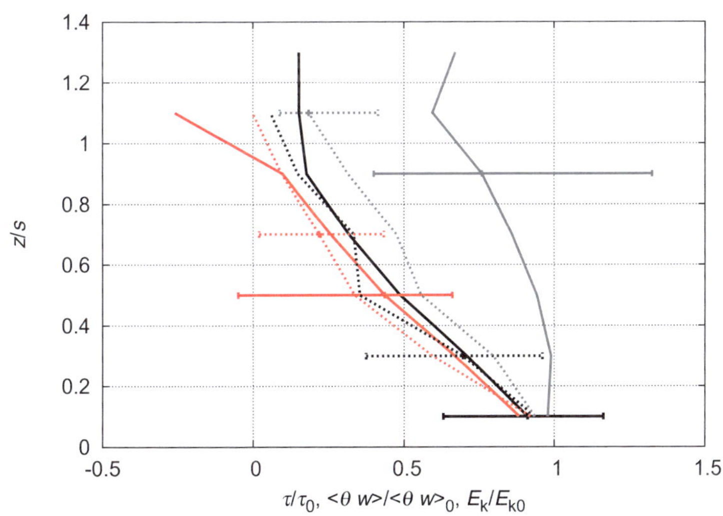

Fig. 3.22 For tSBL: vertical profiles of the momentum flux (*black lines*), the heat flux (*red lines*) and TKE (*gray lines*) normalized over the respective surface values, as function of the height normalized over the vertical variation scale, taken as the average between the momentum and heat flux scales. *Continuous lines*: CCT data; *dashed lines*: SABLES98 data. Some sample *error bars* (one for each profile) suggest the variability of the profiles

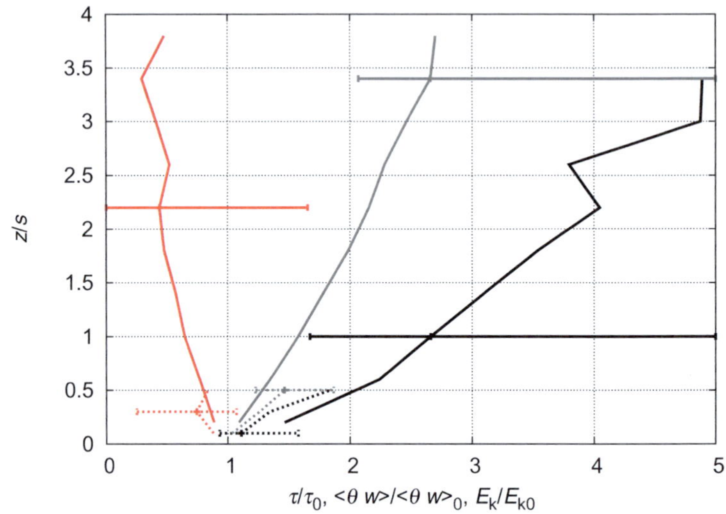

Fig. 3.23 As in Fig. 3.22 but for uSBL. Here the vertical variation scale is taken as the average between the momentum flux and TKE scales

ways; here, the statistical moment y is fitted by $y = a + bz$ and its variation scale is $s = |a/b|$.)

As far tSBL are concerned (Fig. 3.22), the occurrence of median values of the normalized heat flux very near zero indicates that there are cases in which stability (measured by the Obukhov length) changes its sign at the top of SBL: this experimental observation relates to the smallness of the turbulent fluxes and the consequent indeterminacy. The difference between the TKE profiles of the two data sets highlights the variability of the SBL, and the difficulty to reduce the description to a few parameters.

As far uSBL are concerned (Fig. 3.23), the heat flux can be either increasing or decreasing with height, so that the median value is ranging from 1 to 0, hiding the wide variability.

These vertical profiles of second-order moments can be compared with the earlier observations by Caughey et al. (1979, Fig. 5), by Lenschow et al. (1988, Figs. 6, 7, 10), by Sorbjan (1988, Figs. 2, 3, 4), and others. These papers deal with tSBL cases, and the last two with the same data set (with partially different interpretations). Profiles of horizontal velocity variance referring to uSBL cases are reported by Banta et al. (2006, see Figs. 12 and 13d).

3.7.1 Local Similarity

The hypothesis that turbulence is unaffected by the distance from the surface, or from the top of the SBL (the assumption of local similarity) reduces the problem of the formulation of similarity functions to a local problem, that is, allows to substitute the surface value of the Obukhov length L with its local value Λ. Thus, here $\zeta = z/\Lambda$.

3.7.1.1 The Mean Velocity Profile

Let apply the scaling approach to turbulence characterized by flux Richardson number smaller than the critical value. Using the definition (3.9) of R_f (with $V = 0$), the mean velocity gradient can be substituted into (3.38), so the non-dimensional mean velocity gradient reads:

$$\Phi_\mathrm{U} = \frac{\zeta}{R_\mathrm{f}} . \tag{3.103}$$

If the Richardson number reaches its critical (constant) value R_fc, Φ_U turns out to be linear in ζ (and the dimensional mean velocity gradient does not depend on z): this is thus the profile which holds for large ζ and $R_\mathrm{f} \to R_\mathrm{fc}$. A commonly used expression (Högström 1996) matches the quasi-neutral profile and the stable profile:

$$\Phi_\mathrm{U} = 1 + \alpha\zeta . \tag{3.104}$$

The same form is derived by Zilitinkevich et al. (2013). Note that the value of the empirical constant $\alpha = R_\mathrm{fc}^{-1}$ gives an estimate of the critical flux Richardson number. Note also that the independency of the velocity gradient on z supports the validity of the z-less hypothesis. It has been observed by many authors that at increasing ζ over some threshold value, Φ_U increases less than linearly or even becomes almost constant (see, for instance Kondo et al. 1978; Cheng and Brutsaert 2005; Yagüe et al. 2006), and in general the spread of the data increases: see Fig. 3.24.

An empirical approach, frequently used in practical applications (numerical models of atmospheric circulation for weather and climate, or air quality), is given by the Beljaars and Holtslag (1991) formulation:

$$\Phi_\mathrm{U} = 1 + \alpha\zeta + \beta\zeta\,(1 + \gamma - \delta\zeta)\exp(-\delta\zeta) , \tag{3.105}$$

see Beljaars and Holtslag (1991, their Eqs. (28) and (32), coefficients reported after Eq. (32)); see also Luhar et al. (2009).

The formulas (3.104) and (3.105) are shown in Fig. 3.24; the numerical values of the coefficients are reported in Table 3.5.

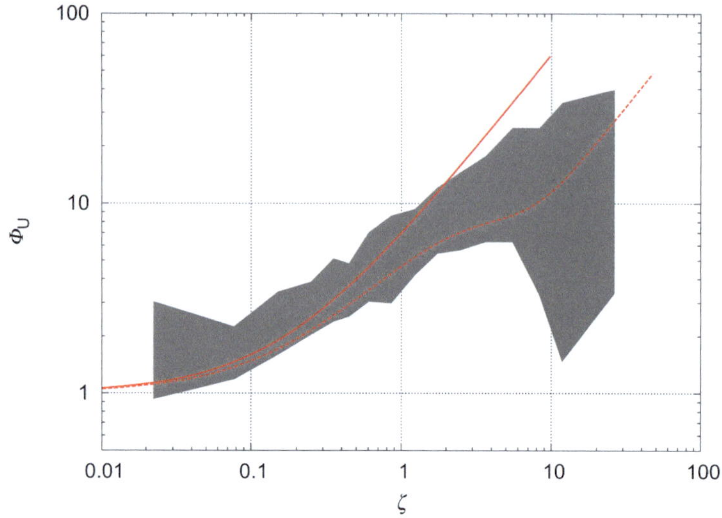

Fig. 3.24 Φ_U as function of ζ. The *gray area*, redrawn from Yagüe et al. (2006), represents the envelope of data from their lowest sonic anemometer. The *full red line* corresponds to (3.104) with Högström (1996) parameter value and the *dashed red line* to (3.105)

Table 3.5 Numerical coefficients for mean wind formulas (stable conditions)

Beljaars and Holtslag (1991)	$\alpha = 1$	$\beta = 2/3$	$\gamma = 5$	$\delta = 0.35$
Högström (1996)	$\alpha = 6$			
Grachev et al. (2005)	$\alpha = 5$			
Zilitinkevich et al. (2013)	$\alpha = 4.0 \div 4.5$			

By recognizing the relevance of the transition from subcritical to supercritical conditions, Grachev et al. (2013) investigated the role of the Richardson number in the choice of the shape of Φ_U. The same point is addressed by Zilitinkevich et al. (2013). It may be observed that for subcritical conditions, (3.104) is the parameterization that better represents the data. It results that Φ_U is smaller for supercritical than for subcritical conditions: in supercritical conditions it can be argued that the correlation between u and w (the momentum flux) includes to a larger extent the contributions from low-frequency phenomena (for instance, undulatory perturbations) that do not contribute to the turbulent mixing, so that u_* turns out to be greater than the velocity scale appropriate for explaining the actual vertical gradient of mean velocity.

The mean velocity profile is obtained by integration in the surface layer (if it exists), or in general in a layer between z_1 and z_2 ($z_2 > z_1$) shallow enough (formally, $(z_2 - z_1) \ll \Lambda[\mathrm{d}\Lambda/\mathrm{d}z]^{-1}$) to neglect variations of Λ. From (3.104) it results:

$$\kappa \frac{U(z_2) - U(z_1)}{u_*} = \ln \frac{z_2}{z_1} + \alpha\,(\zeta_2 - \zeta_1) \ , \qquad (3.106)$$

while from (3.105) it results:

$$\kappa \frac{U(z_2) - U(z_1)}{u_*} = \ln \frac{z_2}{z_1} + \alpha(\zeta_2 - \zeta_1)$$
$$+ \beta \left[(\zeta_2 - \gamma/\delta) \exp(-\delta\zeta_2) - (\zeta_1 - \gamma/\delta) \exp(-\delta\zeta_1) \right] , \qquad (3.107)$$

where Λ is estimated at an height in the (z_1, z_2) interval. These formulas are strictly valid in subcritical conditions.

Some observations are reported in Fig. 3.25: here the data are separated according to the criterion $R_f < 0.25$: subcritical conditions, $R_f > 0.25$: supercritical conditions. It is quite evident that in subcritical conditions the log-linear profile (3.106) gives a fair description of the data.

3.7.1.2 The Mean Temperature Profile

In analogy with the mean velocity gradient, for the mean temperature gradient a linear function is suggested by van Ulden and Holtslag (1985), or by Högström (1996):

$$\Phi_\Theta = \alpha_0 + \alpha_1 \zeta \ . \qquad (3.108)$$

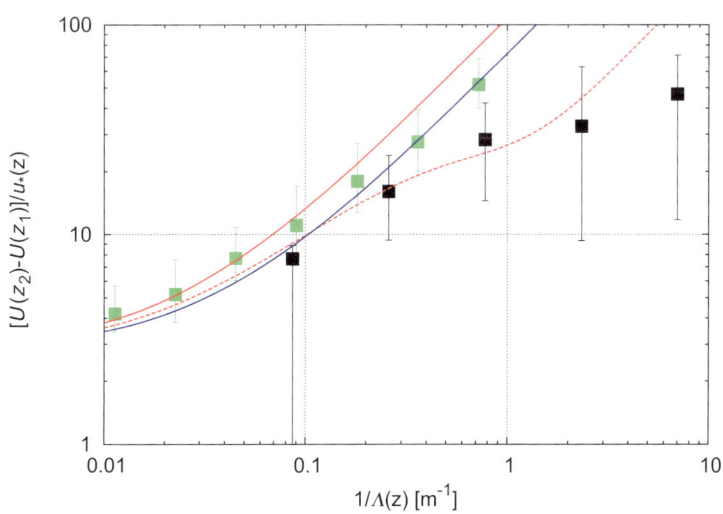

Fig. 3.25 $[U(z_2) - U(z_1)]/u_*$ as function of $1/\Lambda$. Data from SABLES98 are reported with *green symbols* for $R_f < 0.25$ and with *black symbols* for $R_f > 0.25$. Here $z_2 = 10$ m, $z_1 = 3$ m, $z = 5.8$ m. The *full red line* corresponds to (3.106) with Högström (1996) coefficient, the *full blue line* to the same expression with the Zilitinkevich et al. (2013) coefficient and the *dashed red line* to (3.107)

From the integral form by Beljaars and Holtslag (1991) (here Eq. 3.112), the corresponding Φ_Θ can be derived:

$$\Phi_\Theta = 1 + \alpha\zeta\left(1 + \frac{2}{3}\alpha\zeta\right)^{1/2} + \beta\zeta\,(1 + \gamma - \delta\zeta)\exp(-\delta\zeta)\ , \tag{3.109}$$

with the same values of the coefficients given for (3.105).

Zilitinkevich et al. (2013, Eq. (86)) derive a quadratic form for Φ_Θ:

$$\Phi_\Theta = \left(1 + \frac{\beta_1\zeta + \beta_2\zeta^2}{1 + \beta_3\zeta}\right)(1 + \alpha\zeta)\ , \tag{3.110}$$

where the coefficients β_i are given by their Eqs. (83), (84), (85).

The numerical values of the coefficients are reported in Table 3.6.

Some data and the above-cited curves are reported in Fig. 3.26. A behaviour similar to that of the mean velocity for large ζ is observed: here the data suggest that possibly Φ_Θ decreases. The explanation is still related to the influence of

Table 3.6 Numerical coefficients for mean temperature formulas (stable conditions)

van Ulden and Holtslag (1985)	$\alpha_0 = 1$	$\alpha_1 = 5$		
Beljaars and Holtslag (1991)	$\alpha = 1$	$\beta = 2/3$	$\gamma = 5$	$\delta = 0.35$
Högström (1996)	$\alpha_0 = 0.95$	$\alpha_1 = 8$		
Zilitinkevich et al. (2013)	$\beta_1 = 0.45$	$\beta_2 = 1.0$	$\beta_3 = 3.55$	$\alpha = 4.0 \div 4.2$

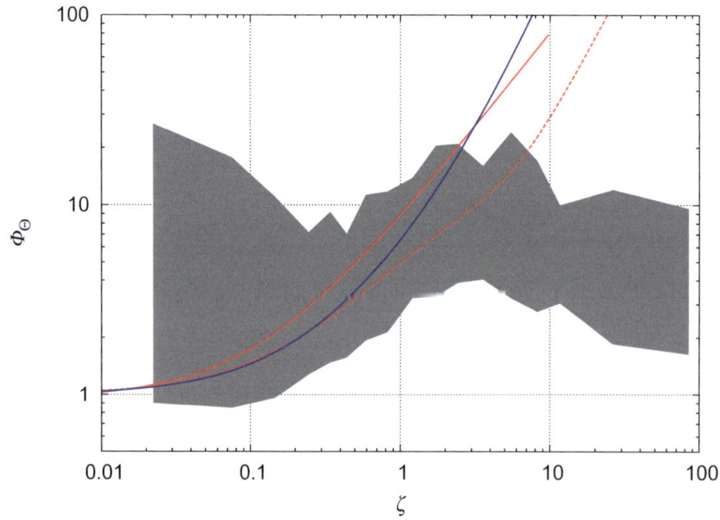

Fig. 3.26 Φ_Θ as function of ζ. The envelope of observations (*gray area*) is redrawn from Yagüe et al. (2006). The *full red line* corresponds to (3.108) with Högström (1996) coefficients, the *dashed red line* to (3.109) and the *full blue line* to (3.110)

low frequency perturbations that increase the correlation between fluctuations but not increase the mixing, i.e. to the presence, in such data, of subcritical and supercritical situations. Moreover it may be observed the large spread for small ζ: for conditions approaching neutrality, the kinematic heat flux becomes negligible and thus θ_* cannot be a proper temperature scale (this point is already addressed in the discussion of the temperature fluctuation variance, see Sect. 3.4.3).

With the same caveats about almost constant fluxes discussed to derive (3.106), from (3.108) the log-linear profile for the mean temperature results:

$$\kappa \frac{\Theta(z_2) - \Theta(z_1)}{-\theta_*} = \alpha_0 \ln \frac{z_2}{z_1} + \alpha_1 (\zeta_2 - \zeta_1) \ . \tag{3.111}$$

Beljaars and Holtslag (1991) suggested an empirical form which partially adjusts to the large ζ behaviour:

$$\begin{aligned} \kappa \ \frac{\Theta(z_2) - \Theta(z_1)}{-\theta_*} &= \ln \frac{z_2}{z_1} \\ &+ \left(1 + \frac{2}{3}\alpha\zeta_2\right)^{3/2} + \beta \left[(\zeta_2 - \gamma/\delta) \exp(-\delta\zeta_2)\right] \\ &- \left(1 + \frac{2}{3}\alpha\zeta_1\right)^{3/2} - \beta \left[(\zeta_1 - \gamma/\delta) \exp(-\delta\zeta_1)\right] \ . \end{aligned} \tag{3.112}$$

Also Eq. (3.110) can be integrated, see Exercise 3.5.

Some observations are reported in Fig. 3.27: the data are separated according to the same criterion used for the mean wind profile ($R_f < 0.25$: subcritical conditions, $R_f > 0.25$: supercritical conditions). It is quite evident that in subcritical conditions the profiles described by the previously-cited formulas give a fair description of the data, while the supercritical behaviour is missed.

3.7.1.3 The Eddy Diffusion Coefficients and the Richardson Numbers

Similarly to Sect. 3.4.2.3, the local value of the eddy diffusion coefficient for momentum in moderately stable conditions, under the limit of critical flux Richardson number, reads:

$$K_\tau = \kappa R_{fc} \Lambda u_* \ , \tag{3.113}$$

which corresponds to the mixing length independent on the distance from the surface:

$$\langle \alpha l^2 \rangle^{1/2} = \kappa R_{fc} \Lambda \ , \tag{3.114}$$

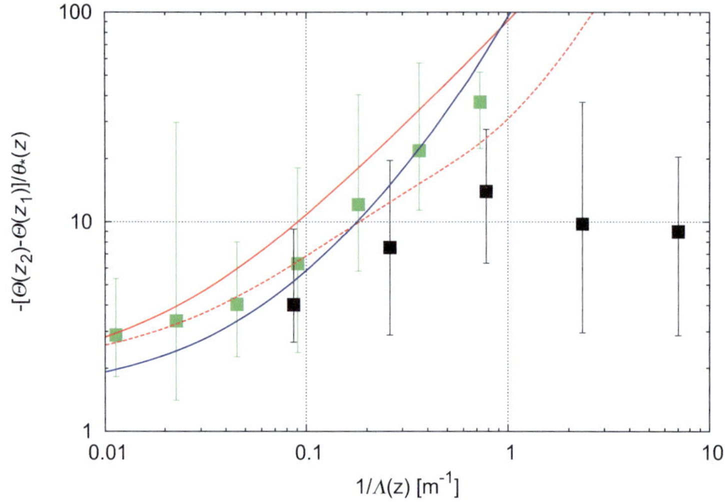

Fig. 3.27 $-[\Theta(z_2) - \Theta(z_1)]/\theta_*$ as function of $1/\Lambda$. Data from SABLES98 experiment are reported with *green symbols* for $R_f < 0.25$ and with *black symbols* for $R_f > 0.25$. Here $z_2 = 8$ m, $z_1 = 3.55$ m and $z = 5.8$ m. The *full red line* corresponds to (3.111) with Högström (1996) coefficients, the *dashed red line* to (3.112), and the *full blue line* to the profile obtained from (3.110): see (S3.26)

at variance with the expression for quasi-neutral and unstable conditions (3.57). These results are consistent with the z-less hypothesis.

The gradient Richardson number and the flux Richardson number can be computed from (3.55) and (3.56) as functions of the Obukhov length using the nondimensional gradients discussed above, and must be tested against data. Also, the turbulent Prandtl number can be derived and tested. The available observations are contradicting (see, for instance Tampieri et al. 2015); in the Fig. 3.28 the relations between R_g and ζ, R_f and P_t are represented, from different data sets and from some parameterizations. It is worth noting that according to the Högström formulas, both R_f and R_g reach a critical value as ζ increases, which is not consistent with the observations. In fact, R_g increases as ζ increases. It is also evident that the different choices of parameterization lead to different turbulent Prandtl numbers.

3.7.2 The Second-Order Moments

In the unstable boundary layer, the ratio of the variances of the velocity on the double of the TKE has been shown early: the variance of the vertical component increases as the Obukhov length decreases in absolute value. The opposite occurs in stable conditions, where the vertical component decreases of almost one order of magnitude as R_g increases from 0.1 to1: see Fig. 3.29.

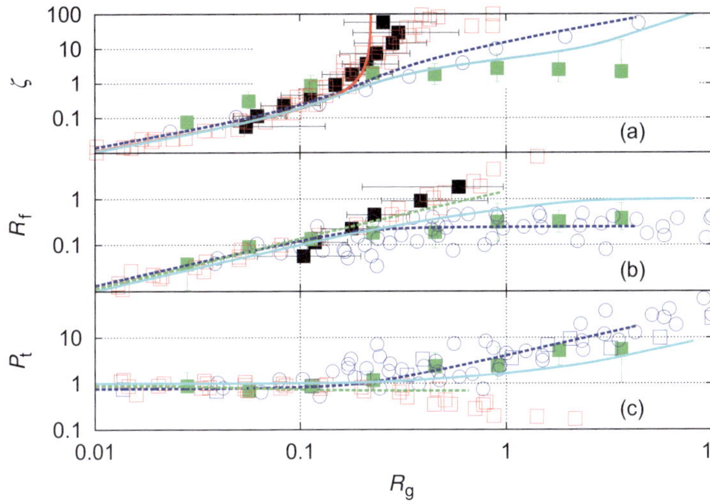

Fig. 3.28 Turbulence parameters as function of R_g. (**a**): ζ. *Squares with error bars*: SABLES98 data, binned for intervals of z/Λ (*black*) and for intervals of R_g (*green*). *Open red squares*: binned data redrawn from Sorbjan and Grachev (2010, Fig. 4). *Blue circles*: binned data redrawn from Zilitinkevich et al. (2013, Fig. 12). *Cyan line*: from Beljaars and Holtslag (1991) formulas for the nondimensional gradients of mean velocity and temperature. *Red line*: from Högström (1996) formulas. *Blue line*: from Zilitinkevich et al. (2013) formulas. For increasing stability the two binning choices diverges, because the scatter of data is quite large: if ζ is considered the independent variable, the Richardson number presents a critical value, consistent with Högström (1996); if R_g is considered the independent variable, the behaviour is more consistent with Zilitinkevich et al. (2013) theoretical curve and data. (**b**): R_f. *Lines and symbols* as before, with the exception of the *open red squares*, from Sorbjan (2010, Fig. 5) and the *gray circles* from Zilitinkevich et al. (2013, Fig. 4). *Green line*: from Sorbjan (2010). (**c**): P_t. *Lines and symbols* as before; but *open red squares*: binned data redrawn from Sorbjan and Grachev (2010, Fig. 2); *blue circles* (atmospheric obs.) and *squares* (numerical simulations): binned data redrawn from Zilitinkevich et al. (2013, Fig. 5)

Furthermore, the behaviour of τ/E_K, $A_1 = \langle\theta w\rangle^2/[E_k\langle\theta^2\rangle]$ and $A_2 = E_P/[E_k + E_P$ give us further indications about the turbulence features in stable conditions: see Fig. 3.30. It results that the ratio of the momentum flux with TKE decreases for increasing stability, i.e. at $R_g > 1$ the stable boundary layer maintains some TKE while the momentum flux decreases. A similar behaviour is shown by the heat flux. Finally, the TPE increases with respect to the total mean energy of the system, as R_g increases.

Note that it is common in the literature to represent the velocity variances normalized to the friction velocity, and also as a function of z/Λ, in a similar way as in the case of unstable conditions. The picture that derives shows that the normalized velocity variances increase with stability, according to the observation that some TKE survives when the momentum flux decreases, but the effect of the self-correlation between $\langle u_i u_i\rangle/u_*^2$ and z/Λ may be dominant for a quantitative determination at $z/\Lambda \gg 1$, as discussed in Sect. 3.3.

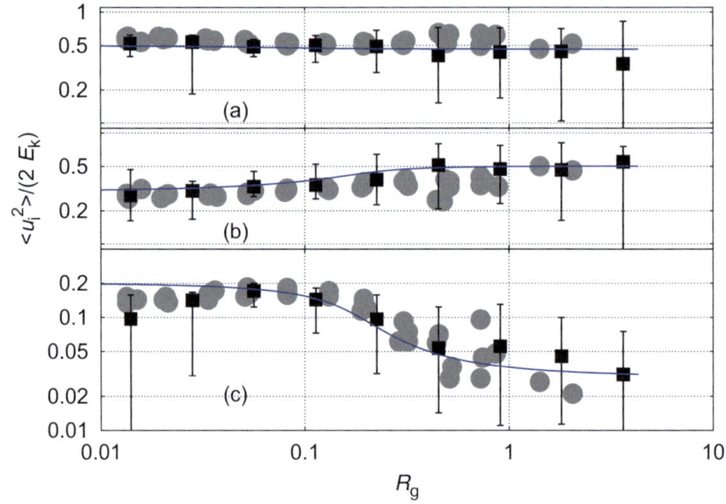

Fig. 3.29 Ratios of the wind component variances on the double of the TKE in the SBL. (**a**): u component. (**b**): v component. (**c**): w component. *Black symbols*: SABLES98 observations. *Gray symbols*: binned observations redrawn from Sorbjan (2010, Fig.7). *Blue line*: from Zilitinkevich et al. (2013), revised coefficients

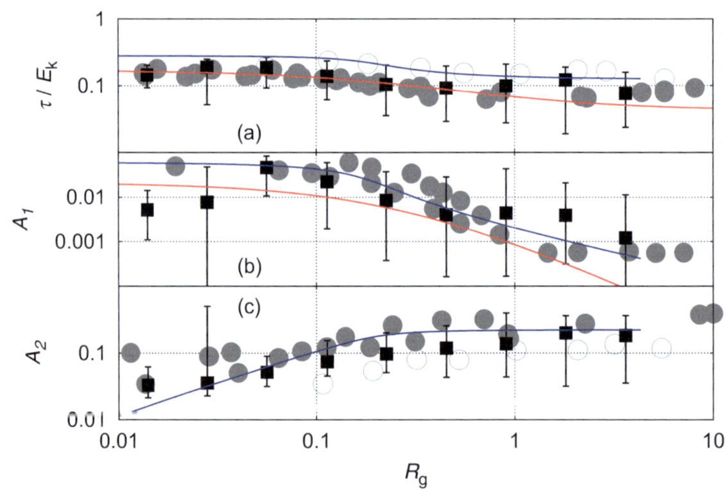

Fig. 3.30 (**a**): τ/E_K as function of R_g. (**b**): $A_1 = \langle\theta w\rangle^2/[E_K\langle\theta^2\rangle]$. (**c**): $E_P/[E_K + E_P]$. *Black symbols*: SABLES98 observations. *Gray full symbols*: atmospheric observations, *open symbols*: laboratory observations, from Zilitinkevich et al. (2013, Figs. 8, 9 and 7 of the revised version). *Red line*: Mauritsen and Svensson (2007). *Blue line*: Zilitinkevich et al. (2013), revised coefficients

3.7.3 The Nieuwstadt (1984) Model

A prototype of tSBL is given by Nieuwstadt (1984, 1985) model (revised for instance by Derbyshire 1990). This model is based on the assumption of Richardson numbers constant with height, and suggests that turbulence is characterized by local scales (z-less assumption).

From (3.3) for the mean temperature and from (3.4) for the mean momentum, neglecting viscosity effects, we can derive the evolution equations under the hypothesis of steady vertical gradients of the mean velocity and the mean temperature (for instance, $\partial^2\Theta/\partial z\partial t = 0$). Thus the model equations turn out to be:

$$\frac{\mathrm{d}^2\langle\theta w\rangle}{\mathrm{d}z^2} = 0 \tag{3.115}$$

$$-\imath f\frac{\mathrm{d}\mathbf{V}}{\mathrm{d}z} + \frac{\mathrm{d}^2\mathbf{X}}{\mathrm{d}z^2} = 0 \tag{3.116}$$

The closure of the model arises from the assumption that the flux Richardson number R_f and the gradient Richardson number R_g are both constant and equal to their critical values, R_fc and R_gc. In the original paper, this value was set to 0.2.

The boundary conditions for (3.115) and (3.116) are:

$$\frac{\langle\theta w\rangle}{\langle\theta w\rangle_0} = 1\ , \quad \frac{|\mathbf{X}|}{u_*^2} = 1 \text{ for } \frac{z}{h} = 0\ , \tag{3.117}$$

$$\frac{\langle\theta w\rangle}{\langle\theta w\rangle_0} = 0\ , \quad \frac{|\mathbf{X}|}{u_*^2} = 0 \text{ for } \frac{z}{h} = 1\ , \tag{3.118}$$

$$\mathbf{V} = 0 \text{ for } \frac{z}{h} = 0\ , \tag{3.119}$$

$$\mathbf{V} = \mathbf{V}_\mathrm{g} \text{ for } \frac{z}{h} = 1\ , \tag{3.120}$$

where h is the boundary layer depth.

To derive the solution, see the Exercise 3.6.

Equation (3.115) has the linear solution:

$$\langle\theta w\rangle = \langle\theta w\rangle_0\left(1 - \frac{z}{h}\right)\ . \tag{3.121}$$

The solution for the momentum flux reads:

$$\mathbf{X} = (1 - z/h)^{3/2}\exp\left(\imath\frac{\sqrt{3}}{2}\log(1 - z/h)\right)\ , \tag{3.122}$$

leading to the power-law rule for the modulus $\tau = |\mathbf{X}|$:

$$\frac{\tau}{u_*^2} = \left(1 - \frac{z}{h}\right)^{3/2} . \tag{3.123}$$

To obtain this solution, an expression for the depth h in terms of the other model parameters results:

$$h = \gamma_c \left(\frac{u_* L}{f}\right)^{1/2} \tag{3.124}$$

being $\gamma_c = (\sqrt{3}\kappa R_{fc})^{1/2} \approx 0.4$ and L is defined from the surface values of the fluxes. This relation may be compared with the expression (3.16) for the neutral case.

The vertical gradient of the mean velocity reads:

$$\frac{L}{u_*}\frac{\mathrm{d}\mathbf{V}}{\mathrm{d}z} = \frac{1}{\kappa R_{fc}}\left(1 - \frac{z}{h}\right)^{(-1+\imath\sqrt{3})/2} , \tag{3.125}$$

and the vertical gradient of the mean temperature reads:

$$-\frac{L}{\theta_*}\frac{\mathrm{d}\Theta}{\mathrm{d}z} = \frac{R_{gc}}{\kappa R_{fc}^2}\left(1 - \frac{z}{h}\right)^{-1} . \tag{3.126}$$

The vertical profile of mean wind is obtained by integrating (3.125). It results:

$$\frac{\mathbf{V}_g - \mathbf{V}(z)}{|\mathbf{V}_g|} = \mathrm{e}^{-\frac{\imath\pi}{3}}\left(1 - \frac{z}{h}\right)^{\frac{1}{2}(1+\imath\sqrt{3})} , \tag{3.127}$$

where the geostrophic wind is:

$$\mathbf{V}_g = |\mathbf{V}_g|\mathrm{e}^{-\frac{\imath\pi}{3}} , \; |\mathbf{V}_g| = \frac{u_* h}{L\kappa R_{fc}} . \tag{3.128}$$

Note that (3.128) together with (3.124) show that in the frame of this model the geostrophic wind, the boundary layer height and the friction velocity are strictly related, as functions of stability and Coriolis parameter. In other words, if for instance the geostrophic wind and the momentum flux at the surface are specified, all the remaining quantities are defined. For the vertical profile of mean velocity an Ekman-like spiral results.

For $z/h \ll 1$, (3.125) gives

$$\mathbf{V}(z) = \frac{u_*}{\kappa R_{fc}}\frac{z}{L} , \tag{3.129}$$

and thus the v component results zero, whereas the u component presents a linear profile, in agreement with the similarity theory for critical flux Richardson number, see (3.103).

Note that (3.127) gives a supergeostrophic velocity at heights of the order of $0.9h$ (a manifestation of 'low-level jet') and that the geostrophic velocity is the correct scale for the wind in SBL (excluding obviously the surface layer). Derbyshire (1990) noted that the wind modulus at the SBL top, (3.128), is equal to the extrapolation to $z = h$ for the near-ground linear profile (3.129).

The vertical profile of mean temperature is obtained by integrating (3.126). It results:

$$\Theta(z) = \Theta(0) + \frac{\theta_* h}{L} \frac{R_{\mathrm{gc}}}{\kappa R_{\mathrm{fc}}^2} \ln\left(1 - \frac{z}{h}\right), z < h \,. \tag{3.130}$$

For $z/h \ll 1$ the expansion of (3.130) gives the temperature profile near the ground:

$$\Theta(z) = \Theta(0) - \frac{\theta_* R_{\mathrm{gc}}}{\kappa R_{\mathrm{fc}}^2} \frac{z}{L} \,. \tag{3.131}$$

The eddy diffusion coefficients consistent with this model are derived using the flux-gradient relationship:

$$K_\tau = u_* \kappa L R_{\mathrm{fc}} \left(1 - \frac{z}{h}\right)^2, K_{\mathrm{Q}} = K_\tau \frac{R_{\mathrm{fc}}}{R_{\mathrm{gc}}} \,, \tag{3.132}$$

and the turbulent Prandtl number turns out to be constant.

The solution is presented in the Figs. 3.31 and 3.32.

Dimensional arguments and the local scaling suggest that the variances of the velocity components and the turbulent kinetic energy are proportional to $\tau(z)$, and that the potential temperature variance is proportional to $\langle \theta w \rangle / \tau^{1/2}$. Thus all the second order moments are expected to become negligible at the same height, the boundary layer depth.

Observations of profiles of wind and of temperature difference can be compared with the Nieuwstadt model result, with some care, as in Fig. 3.33. These data refer to tSBL cases, and are selected in order to have the same h/L ratio of the model (see Exercise 3.7 for more details). In this case, the gradient Richardson number is smaller than the critical value; the mean profiles show near the surface the logarithmic shape, which is not present in the model, and are reasonably well fitted by the surface layer profiles, from Högström (1996).

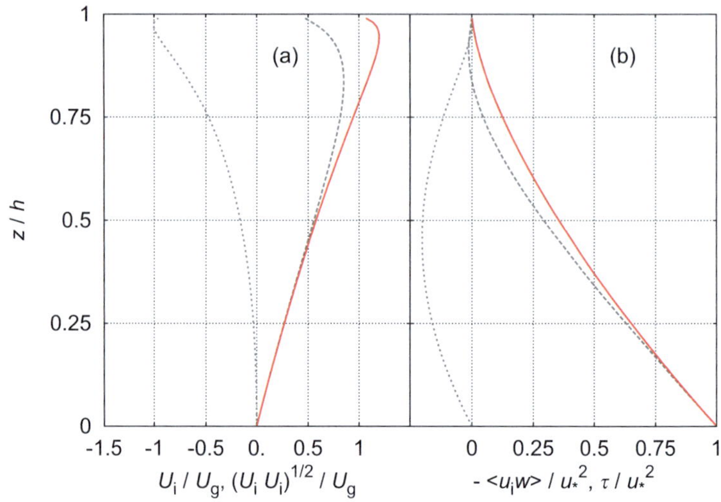

Fig. 3.31 In (**a**): vertical profiles of the horizontal velocity components in direction x (*dashed line*), in direction y (*dotted line*) and of the modulus of the velocity (*continuous red line*) normalized over the modulus of the geostrophic velocity (in this example, the x and y components of the geostrophic velocity are equal). In (**b**): vertical profiles of the negative of the components of the vertical momentum flux (*dashed line*: x component; *dotted line*: y component) and of the modulus of the momentum flux (*continuous red line*), normalized over its surface value u_*^2. The height is normalized with h, from (3.124)

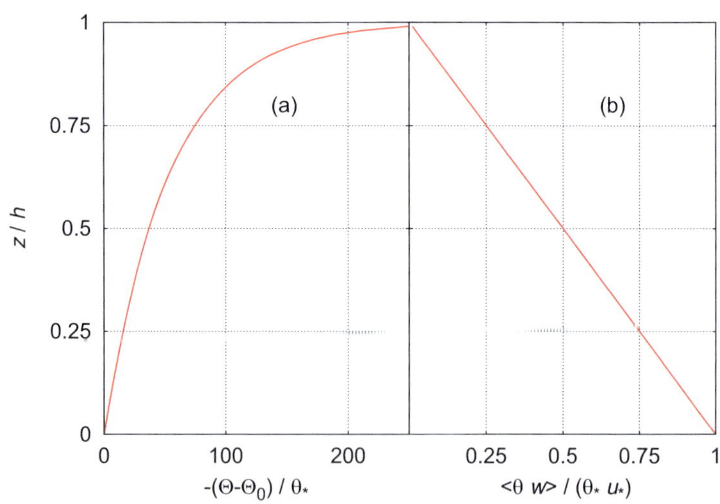

Fig. 3.32 In (**a**): vertical profiles of the mean temperature excess over the surface value, normalized over the negative of the temperature scale $-\theta_*$. In (**b**): vertical profile of the heat flux, normalized over its surface value $\theta_* u_*$. The height is normalized with h, from (3.124)

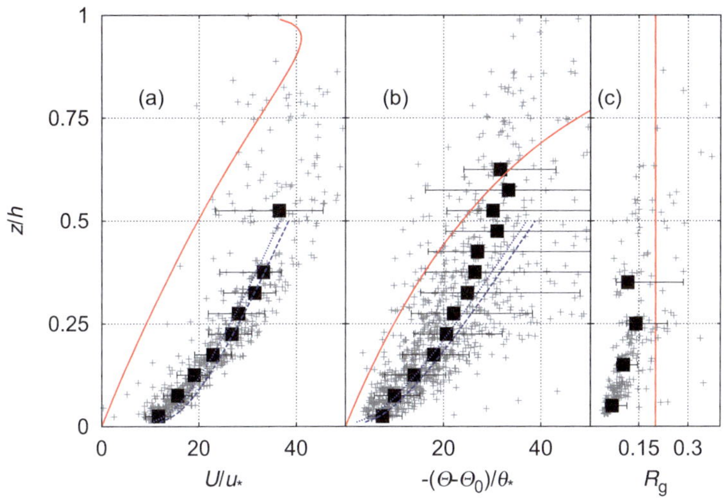

Fig. 3.33 Vertical profiles of mean wind (**a**), mean temperature difference (**b**) and gradient Richardson number (**c**), for tSBL cases, from SABLES98 data. *Red line*: Nieuwstadt model; *blue dashed line*: from Högström (1996), $h = 100\,\text{m}$, *blue dotted line*: the same, with $h = 50\,\text{m}$

3.7.4 The Neutral and Stable Boundary Layer Depth

The stable boundary layers dominated by the surface fluxes lead to identify the depth in a straightforward manner, as the height at which the surface fluxes reduce to a given fraction of their surface values: tSBL corresponds to this description.

In these cases, using the same approach outlined in Sect. 3.5.2, the scale of the mixing length is the Obukhov length L and the eddy diffusion coefficient is $K_\tau \propto u_* L$. Thus, from (3.72), it results:

$$h^2 \propto \frac{u_* L}{f} , \tag{3.133}$$

which is the expression for the 'nocturnal stable' boundary layer (Zilitinkevich 2002; Zilitinkevich et al. 2007), based on the values of the momentum and heat fluxes at the surface.

This expression applies as the background stratification can be considered unimportant ($N \approx 0$). Note that (3.133) has the same functional form as the relation for SBL depth resulting from the Nieuwstadt model, (3.124).

Especially during the wintertime at high latitudes, 'long-lived stable' boundary layers occur, capped by a stably stratified troposphere. For such SBL, the effect of the surface fluxes, accounted for in (3.133), can be superimposed to the effect of the stability aloft, accounted for in (3.75).

The multi-limit expression

$$h^{-2} = \frac{f^2}{C_1^2 \tau_0} + \frac{N|f|}{C_2^2 \tau_0} - \frac{|f|(g/\theta_{00})\langle \theta w \rangle_0}{C_3^2 \tau_0^2}, \tag{3.134}$$

where $C_1 = 0.6$, $C_2 = 1.36$, $C_3 = 0.51$ (Zilitinkevich and Mironov 1996; Zilitinkevich and Esau 2007; Zilitinkevich et al. 2007) is suitable to be used in neutral and stable conditions, and accounts for the effects of surface fluxes, stratification of the atmosphere aloft, and Earth rotation (see also Mironov and Fedorovich 2010).

Estimates based on the mean profiles of velocity and temperature are widely used, and refer to the concept of a critical value of the gradient Richardson number. Clearly, the boundary layer depth can be related to the height at which turbulence is suppressed by stratification, i.e. at which a critical value of the Richardson number is reached. Instead of the gradient Richardson number, the bulk Richardson number (3.21) based on the PBL depth is used:

$$h = R_B(0, h, z_0, h) \frac{\Theta_r U(h)^2}{g[\Theta(h) - \Theta(0)]}. \tag{3.135}$$

The bulk Richardson number is computed from the surface upwards: it is expected to increase, and the PBL depth is found as R_B reaches a critical value to be defined (Mahrt 1981; Troen and Mahrt 1986). Many empirical modifications are available from literature.

The Nieuwstadt solution allows to compute this height for that case. Using (3.127) and (3.130) $R_B(0, z, z_0, z)$ turns out to be function of z/h, it increases from the surface to the top, and is independent on the stability at the ground. As discussed by Zilitinkevich and Baklanov (2002), the Richardson-based criteria are not fully reliable; however, it can be a rough estimate of the equilibrium depth in absence of detailed flux measurements.

3.8 Some Remarks About the Spectra

In shear-dominated conditions, the spectra are expected to be K41 compliant in the large wavenumber, nearly isotropic inertial subrange, while the large scales are affected by the boundary conditions and by the meteorological influences. A classical example is reported in Fig. 3.34. A recent discussion is given by Larsén et al. (2016).

In unstable conditions, the spectra are expected to present an influence of the PBL depth, which is the scale of the largest eddies (Yaglom 1994).

Some wind velocity spectra for unstable conditions are shown in Fig. 3.35. Note that differences of stability affect more the vertical component of velocity than the horizontal one, and that the horizontal and vertical components differ at the

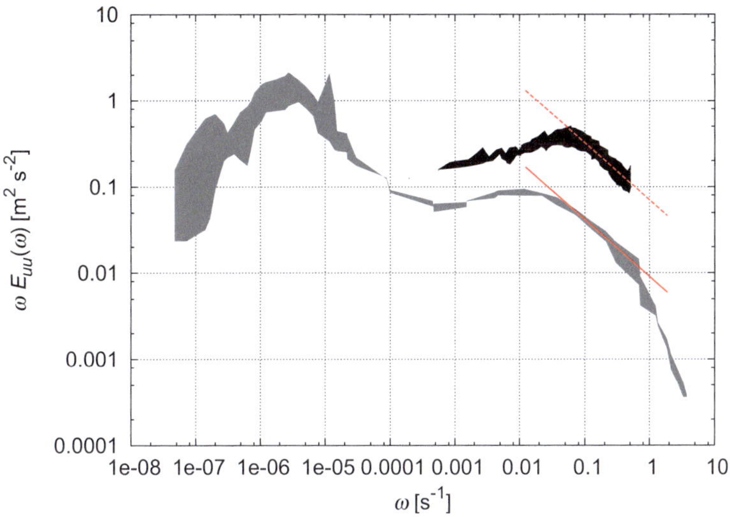

Fig. 3.34 Energy spectra, derived applying the frozen turbulence hypothesis, redrawn from Courtney and Troen (1990) (*gray*) and from Anfossi et al. (2005) (*black*). The data from Courtney and Troen (1990) refer to a 1-year long record, and show not only the inertial subrange but also the typical low-frequency behaviour related to the weather systems. The gap allows to separate PBL turbulence from the large-scale turbulence (which often is not considered as such, because of its explicit representation in the meteorological models). Data from Anfossi et al. (2005) refer to measurements made in the surface layer. The TKE associated with the high frequency range is different (*black* data refer to neutral conditions with wind velocity larger than $3\,\mathrm{m\,s^{-1}}$; *gray data* include various conditions), but in both the cases the inertial subrange can be recognized (the *red lines*)

large scales, while the inertial subrange is K41 compliant: it is fairly isotropic and reproduces the slope $\omega^{-5/3}$ as in (2.109).

It is worth noting that the normalized spectrum for the horizontal velocity component presents a wide transition region, between the inertial subrange (where the spectrum increases as the frequency decreases) and the large scale range (where it decreases): in this region, the normalized spectrum is broadly constant so that it is consistent with $E_{\mathrm{uu}}(\omega) \propto \omega^{-1}$, or with $E_{\mathrm{uu}}(k) \propto k^{-1}$. This k^{-1} range is related (Högström et al. 2002) to the existence of large scale longitudinal eddies that affect mainly the horizontal components of the velocity. This large-scale contribution to the horizontal fraction of TKE is independent from the distance to the surface (but not the amplitude of this subrange, being limited by $kz \ll 1$); dimensional arguments thus lead to the k^{-1} power law. This range has been discussed in the frame of wall turbulence by Nikora (1999) for instance; and found in atmospheric observations in different conditions (Mammarella et al. 2005).

Spectra of scalars (mixing ratios of water vapour and CO2) are reported in Fig. 3.36: it is evident the collapse in the inertial subrange.

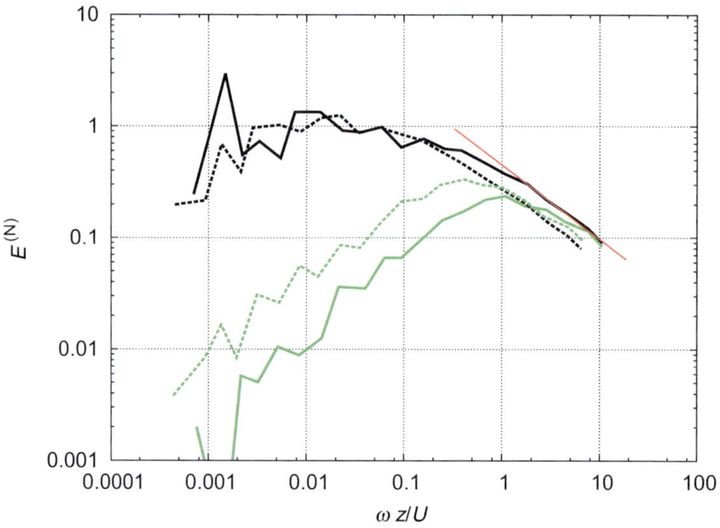

Fig. 3.35 Spectra of velocity components in unstable conditions, normalized according to the inertial subrange scaling (2.109) for velocity: i.e. $E^{(N)} = \omega E_{ii}(\omega)/[z^{2/3}\varepsilon^{2/3}]$. The abscissa is the nondimensional frequency. *Continuous lines*: $\zeta = 0.1$, *dashed lines*: $\zeta = 0.026$. *Black*: *u* component; *green*: *w* component. Redrawn from Smedman et al. (2007, Figs. 6 and 8)

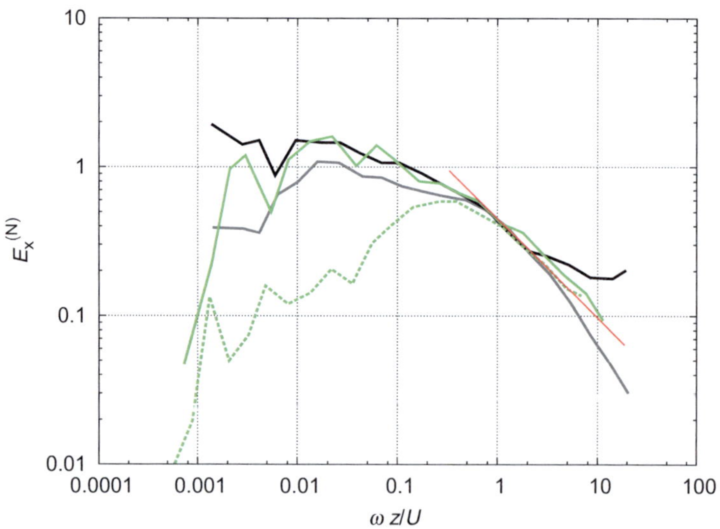

Fig. 3.36 Spectra of scalars in unstable conditions, normalized according to the inertial subrange scaling (2.116) for a scalar *x*: i.e. $E_x^{(N)} = \omega E_x(\omega)/[z^{2/3}\varepsilon^{-1/3}\varepsilon_x]$ remember (2.106). The abscissa is the nondimensional frequency. *Green lines*: temperature: *continuous* for $\zeta = 0.11$, *dashed* for $\zeta = 0.026$, redrawn from Smedman et al. (2007, Figs. 7 and 8). *Gray line*: water vapour; *black line*: CO2, for an average value $\zeta = 0.26$, redrawn from Sahlee et al. (2008, Fig. 4). *Red line*: slope from the inertial subrange spectrum (2.116)

In the stably stratified atmosphere, the stratification induces anisotropy and modifies the spectra of the kinetic energy and of the scalar, because they will depend not only on dissipation, but also on the buoyancy (Bolgiano 1959; Monin and Yaglom 1975).

The wavenumber scale separating the buoyancy-affected range from the shear-dominated range (the inertial subrange) may be evaluated by dimensional analysis: it must depend on the buoyancy production term in the temperature variance equation $\langle\theta w\rangle\, \mathrm{d}\Theta_\mathrm{f}/\,\mathrm{d}z$, on the buoyancy parameter g/θ_{00} and on the dissipation rate ε of TKE:

$$k_\mathrm{B} \propto \left[\langle\theta w\rangle \frac{\mathrm{d}\Theta_\mathrm{f}}{\mathrm{d}z}\right]^{3/4} \left(\frac{g}{\theta_{00}}\right)^{3/2} \varepsilon^{-5/4} = \left[\frac{g}{\theta_{00}}\langle\theta w\rangle\right]^{3/4} N^{3/2}\varepsilon^{-5/4} \,. \qquad (3.136)$$

This wavenumber scale must be compared with the largest wavenumber of the inertial subrange $\mathrm{l}_\mathrm{K}^{-1}$, from (2.92). Being the viscosity and the dissipation fixed, the increase of stability, measured by N, increases the value of k_B while $\mathrm{l}_\mathrm{K}^{-1}$ is constant: it is conceivable that the inertial subrange can disappear for enough large stability conditions, and the whole structure of the small-scale turbulence becomes anisotropic (Bolgiano 1962). The wavenumber range between the value corresponding to the integral length scale and k_B is called the buoyancy subrange.

Thus, in this subrange, the spectra cannot be function of the TKE dissipation, and read:

$$E_\mathrm{uu} \propto \left[\frac{\mathrm{d}\theta_\mathrm{f}}{\mathrm{d}z}\langle\theta w\rangle\right]^{2/5} \left[\frac{g}{\theta_{00}}\right]^{4/5} k^{-11/5} = \left[\frac{g}{\theta_{00}}\langle\theta w\rangle\right]^{2/5} N^{1/5}k^{-11/5} \,, \qquad (3.137)$$

for the velocity variances and for the TKE, and

$$E_{\theta\theta} \propto \left[\frac{\mathrm{d}\theta_\mathrm{f}}{\mathrm{d}z}\langle\theta w\rangle\right]^{4/5} \left[\frac{g}{\theta_{00}}\right]^{-2/5} k^{-7/5} \,, \qquad (3.138)$$

for the temperature variance.

Further, a stably stratified atmosphere, characterized by the Brunt-Väisälä frequency N, leads to the occurrence of gravity waves. In a spectral range where the gravity waves are dominant, the energy spectrum reads:

$$E_\mathrm{uu} \propto N^2 k^{-3} \,, \qquad (3.139)$$

on the basis of dimensional arguments (Finnigan et al. 1984).

Some examples for velocity are reported in Fig. 3.37. Some examples for a scalar (the CO2 mixing ratio) are reported in Fig. 3.38. Note the effect of increasing stability.

To summarize this discussion, the changes in the large scales at varying stability are evidenced in Fig. 3.39, for the temperature spectrum.

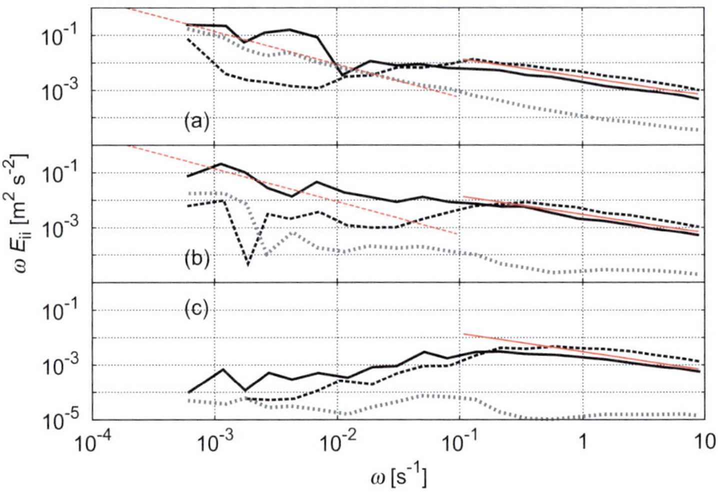

Fig. 3.37 Spectra of wind component variances in the SBL, redrawn from Smedman et al. (1997). (**a**): u component. (**b**): v component. (**c**): w component. *Black continuous and dashed lines*: two cases of intense turbulence; *gray dotted line*: a case of weak turbulence. The gap (not always present) separates the inertial subrange and the buoyancy subrange. The buoyancy subrange hides almost completely the inertial subrange for u component in the low turbulence case. The inertial subrange is reasonably isotropic (the *continuous red line* highlights that the dissipation is the same for the three components, in cases of intense turbulence). *Continuous red line*: slope from the inertial subrange spectrum (2.109); *dashed*: slope from the buoyancy subrange spectrum (3.137)

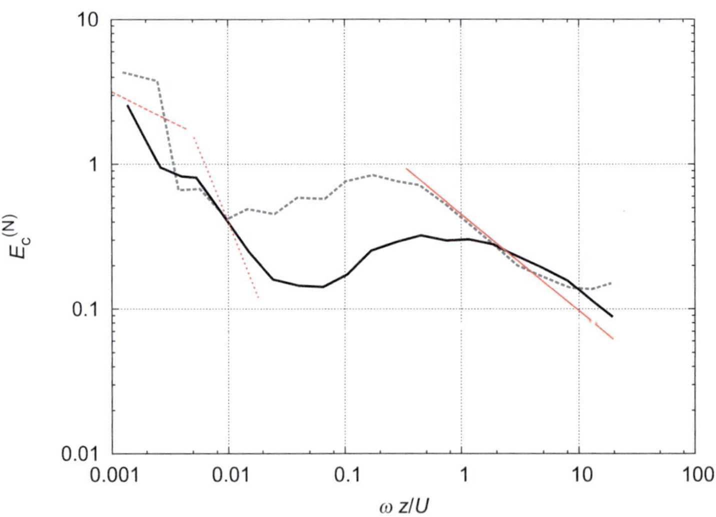

Fig. 3.38 Spectra, normalized as in Fig. 3.36, of the CO2 mixing ratio in the SBL, redrawn from Sahlee et al. (2008). *Gray line*: $0 < \zeta < 0.1$, *black line*: $\zeta > 0.3$. The gap separates the inertial subrange and the buoyancy subrange. *Continuous red line*: inertial subrange spectrum (2.116); *dashed*: buoyancy subrange spectrum (3.138), *dotted*: wave-dominated regime (3.139)

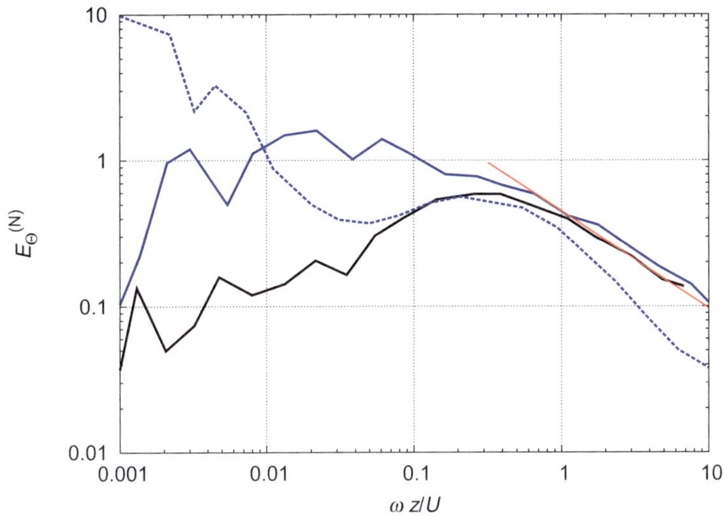

Fig. 3.39 Spectra, normalized as in Fig. 3.36 of the temperature variance for a near neutral case (*black line*), an unstable case (*blue continuous*) and a strong stable case (*blue dashed*), redrawn from Sahlee et al. (2008) and from Smedman et al. (2007)

Exercises

3.1 Derive the expression for the mean turbulent potential energy.

3.2 Derive the mean wind and the mean temperature profiles according to the stepwise (three sublayers) formulation for unstable conditions by Kader and Yaglom (1990).

3.3 Derive the mean wind profile for unstable conditions according to Frenzen and Vogel (2001).

3.4 Derive the Ekman solution and discuss the relations among geostrophic wind, friction velocity and boundary layer depth.

3.5 Derive the mean temperature profile for stable conditions according to the non-dimensional gradient (3.110), proposed by Zilitinkevich et al. (2013).

3.6 Derive the solution of the model by Nieuwstadt (1984).

3.7 Outline a procedure to make the comparison between tower data and Nieuwstadt model, as in Fig. 3.33.

References

J.D. Albertson, M. Parlange, G. Kiely, W.E. Eichinger, The average dissipation rate of turbulent kinetic energy in the neutral and unstable atmospheric surface layer. J. Geophys. Res. **102**, 13423–13432 (1997)

E.L. Andreas, R. Hill, J. Gosz, Statistics of surface-layer turbulence over terrain with metre-scale heterogeneity. Bound.-Layer Meteorol. **86**, 379–408 (1998)

E.L. Andreas, K.J. Claffey, R.E. Jordan, C.W. Fairall, P.S. Guest, P.O.G. Persson, A.A. Grachev, Evaluation of the von Karman constant in the atmospheric surface layer. J. Fluid Mech. **559**, 117–149 (2006)

D. Anfossi, G. Schayes, G. Degrazia, A. Goulart, Atmospheric turbulence decay during the solar total eclipse of 11 August 1999. Bound.-Layer Meteorol. **111**, 301–311 (2004)

D. Anfossi, D. Oettl, G. Degrazia, E. Ferrero, A. Goulart, An analysis of sonic anemometer observations in low wind speed conditions. Bound.-Layer Meteorol. **114**, 179–203 (2005)

W.M. Angevine, A.W. Grimsdell, L.M. Hartten, A.C. Delany, The flatland boundary layer experiments. Bull. Am. Meteorol. Soc. **79**, 419–431 (1998)

A. Ansmann, J. Fruntke, R. Engelmann, Updraft and downdraft characterization with Doppler lidar: cloud-free versus cumuli-topped mixed layer. Atmos. Chem. Phys. **10**, 7845–7858 (2010)

P. Baas, G.J. Steeneveld, B.J.H. van de Wiel, A.A.M. Holtslag, Exploring self-correlation in fluxgradient relationships for stably stratified conditions. J. Atmos. Sci. **63**, 3045–3054 (2006)

D. Bala Subrahamanyam, T.J. Anurose, M. Mohan, M. Santosh, N.V.P. Kiran Kumar, S. Sijikumar, S.S. Prijith, M. Aloysius, Atmospheric surface-layer response to the annular solar eclipse of 15 January 2010 over Thiruvananthapuram, India. Bound.-Layer Meteorol. **141**, 325–332 (2011)

R.M. Banta, Y.L. Pichugina, W.A. Brewer, Turbulent velocity-variance profiles in the stable boundary layer generated by a nocturnal low-level jet. J. Atmos. Sci. **63**, 2700–2719 (2006)

E. Barberis, Analisi statistiche nello strato limite turbolento. Thesis, Univ. Torino, Dip. Fisica (2007)

E. Batchvarova, S.E. Gryning, Applied model for the growth of the daytime mixed layer. Bound.-Layer Meteorol. **56**, 261–274 (1990)

A. Beljaars, A.A.M. Holtslag, Flux parameterization over land surfaces for atmospheric models. J. Appl. Meteorol. **30**, 327–341 (1991)

F. Beyrich, H.T. Mengelkamp, Evaporation over a heterogeneous land surface: EVA_GRIPS and the LITFASS-2003 experiment. Bound.-Layer Meteorol. **121**, 5–32 (2006)

R. Bolgiano, Turbulent spectra in a stably stratified atmosphere. J. Geophys. Res. **64**, 2226–2229 (1959)

R. Bolgiano, Structure of turbulence in stratified media. J. Geophys. Res. **67**, 3015–3023 (1962)

S.J. Caughey, J.C. Kaimal, Vertical heat flux in the convective boundary layer. Q. J. R. Meteorol. Soc. **103**, 811–815 (1977)

S.J. Caughey, J.C. Wyngaard, J.C. Kaimal, Turbulence in the evolving stable boundary layer. J. Atmos. Sci. **36**, 1041–1052 (1979)

G. Caulliez, V.K. Makin, V. Kudryavtsev, Drag of the water surface at very short fetches: observations and modeling. J. Phys. Oceanogr. **38**, 2038–2055 (2008)

H. Charnock, Wind stress on a water surface. Q. J. R. Meteorol. Soc. **81**, 639–640 (1955)

D. Charuchittipan, J.D. Wilson, Turbulent kinetic energy dissipation in the surface layer. Bound.-Layer Meteorol. **132**, 193–204 (2009)

Y. Cheng, W. Brutsaert, Fluxprofile relationships for wind speed and temperature in the stable atmospheric boundary layer. Bound.-Layer Meteorol. **114**, 519–538 (2005)

P. Chiba, Stability dependence of the vertical velocity skewness in the atmospheric surface layer. J. Meteorol. Soc. Jpn. **56**, 140–142 (1978)

M. Courtney, I. Troen, Wind speed spectrum from one year of continuous 8 hz measurements, in *Proceedings 9th Symposium on Turbulence and Diffusion* (American Meteorological Society, Boston, MA, 1990)

C. Darbieu, M. Lothon, D. Pino, Turbulence vertical structure of the boundary layer during the afternoon transition. Atmos. Chem. Phys. **15**, 10071–10086 (2015)
M. de Franceschi, D. Zardi, M. Tagliazucca, F. Tampieri, Analysis of second order moments in surface layer turbulence in an alpine valley. Q. J. R. Meteorol. Soc. **135**, 1750–1765 (2009)
J.W. Deardorff, The counter-gradient heat flux in the atmosphere and in the laboratory. J. Atmos. Sci. **23**, 503–506 (1966)
J.W. Deardorff, Convective velocity and temperature scales for the unstable planetary boundary layer. J. Atmos. Sci. **27**, 1211–1213 (1970)
S.H. Derbyshire, Nieuwstadt's stable boundary layer revisited. Q. J. R. Meteorol. Soc. **116**, 127–158 (1990)
N.L. Dias, W. Brutsaert, M.L. Wesely, Z-less stratification under stable conditions. Bound.-Layer Meteorol. **75**, 175–187 (1995)
F. Durst, J. Jovanovic, L.J. Kanevce, Probability density distribution in turbulent wall boundary-layer flows, in *Turbulent Shear Flows 5*, ed. by F. Durst, B.E. Launder, J.L. Lumley, F.W. Schmidt, J.H. Whitelaw (Springer, Berlin, 1987)
H.J.S. Fernando, B. Verhoef, S. Di Sabatino, L.S. Leo, S. Park, The phoenix evening transition flow experiment (TRANSFLEX). Bound.-Layer Meteorol. **147**, 443–468 (2013)
J.J. Finnigan, F. Einaudi, D. Fua, The interaction between an internal gravity wave and turbulence in the stably stratified nocturnal boundary layer. J. Atmos. Sci. **41**, 2409–2436 (1984)
P. Frenzen, C.A. Vogel, Further studies of atmospheric turbulence in layers near the surface: scaling the tke budget above the roughness sublayer. Bound.-Layer Meteorol. **99**, 173–206 (2001)
J.R. Garratt, *The Atmospheric Boundary Layer* (Cambridge University Press, Cambridge, 1992), 316 pp
A. Goulart, G. Degrazia, U. Rizza, D. Anfossi, A theoretical model for the study of the convective turbulence decay and comparison with les data. Bound.-Layer Meteorol. **107**, 143–155 (2003)
A.A. Grachev, C.W. Fairall, P.O.G. Persson, E.L. Andreas, P.S. Guest, Stable boundary-layer scaling regimes: the SHEBA data. Bound.-Layer Meteorol. **116**, 201–235 (2005)
A.A. Grachev, E.L. Andreas, C.W. Fairall, P.S. Guest, P.O.G. Persson, On the turbulent Prandtl number in the stable atmospheric boundary layer. Bound.-Layer Meteorol. **125**, 329–341 (2007)
A.A. Grachev, E.L. Andreas, C.W. Fairall, P.S. Guest, P.O.G. Persson, The critical Richardson number and limits of applicability of local similarity theory in the stable boundary layer. Bound.-Layer Meteorol **147**, 51–82 (2013)
V. Gryanik, J. Hartmann, A turbulence closure for the convective boundary layer based on a two-scale mass-flux approach. J. Atmos. Sci. **59**, 2729–2744 (2002)
B.B. Hicks, W.R. Pendergrass III, C.A. Vogel, R.N. Keener Jr., S.M. Leyton, On the micrometeorology of the Southern Great Plains 1: legacy relationships revisited. Bound.-Layer Meteorol. **151**, 389–405 (2014)
C.W. Higgins, C. Meneveau, M. Parlange, The effect of filter dimension on the subgrid-scale stress, heat flux, and tensor alignements in the atmospheric surface layer. J. Atmos. Ocean. Technol. **24**, 360–375 (2007)
U. Högström, Analysis of turbulence structure in the surface layer with a modified similarity formulation for near neutral conditions. J. Atmos. Sci. **47**, 1949–1972 (1990)
U. Högström, Review of some basic characteristics of the atmospheric surface layer. Bound.-Layer Meteorol. **78**, 215–246 (1996)
U. Högström, J.C.R. Hunt, A.S. Smedman, Theory and measurements for turbulence spectra and variances in the atmospheric neutral surface layer. Bound.-Layer Meteorol. **103**, 101–124 (2002)
J.C.R. Hunt, J.C. Kaimal, J.E. Gaynor, Eddy structure in the convective boundary layer - new measurements and new concepts. Q. J. R. Meteorol. Soc. **114**, 827–858 (1988)
B.A. Kader, A.M. Yaglom, Mean fields and fluctuation moments in unstably stratified turbulent boundary layers. J. Fluid Mech. **212**, 637–662 (1990)
J.C. Kaimal, J.J. Finnigan, *Atmospheric Boundary Layer Flows. Their Structure and Measurement.* (Oxford University Press, Oxford, 1994)

C.L. Klipp, L. Mahrt, Flux-gradient relationship, self-correlation and intermittency in the stable boundary layer. Q. J. R. Meteorol. Soc. **130**, 2087–2103 (2004)

J. Kondo, O. Kanechika, N. Yasuda, Heat and momentum transfers under strong stability in the atmospheric surface layer. J. Atmos. Sci. **35**, 1012–1021 (1978)

X.G. Larsén, S.E. Larsen, E.L. Petersen, Full-scale spectrum of boundary-layer winds. Bound.-Layer Meteorol. **159**, 349–371 (2016)

D.H. Lenschow, J.C. Wyngaard, W.T. Pennel, Mean field and second moment budgets in a baroclinic, convective boundary layer. J. Atmos. Sci. **37**, 1313–1326 (1980)

D.H. Lenschow, X.S. Li, C.J. Zhu, B.B. Stankov, The stably stratified boundary layer over the Great Plains. Bound.-Layer Meteorol. **42**, 95–121 (1988)

D.H. Lenschow, M. Lothon, S.D. Mayor, P.P. Sullivan, G. Canut, A comparison of higher-order vertical velocity moments in the convective boundary layer from lidar with in situ measurements and large-eddy simulation. Bound.-Layer Meteorol. **143**, 107–123 (2012)

D. Li, G. Katul, E. Bou-Zeid, Mean velocity and temperature profiles in a sheared diabatic turbulent boundary layer. Phys. Fluids **24** (2012). 105105.1–105105.16

J. Liang, L. Zhang, Y. Wang, X. Cao, Q. Zhang, H. Wang, B. Zhang, Turbulence regimes and the validity of similarity theory in the stable boundary layer over complex terrain of the Loess Plateau, China. J. Geophys. Res.-Atmos. **119**, 6009–6021 (2014)

M. Lothon, and B. team, The BLLAST field experiment: boundary-Layer late afternoon and sunset turbulence. Atmos. Chem. Phys. **14**, 10931–10960 (2014)

A.K. Luhar, An analytical slab model for the growth of the coastal thermal internal boundary layer under near-neutral onshore flowconditions. Bound.-Layer Meteorol. **88**, 103–120 (1998)

A.K. Luhar, P.J. Hurley, K.N. Rayner, Modelling near-surface low winds over land under stable conditions: sensitivity tests, flux-gradient relationships, and stability parameters. Bound.-Layer Meteorol. **130**, 249–274 (2009)

L. Mahrt, Modelling the depth of the stable boundary layer. Bound.-Layer Meteorol. **21**, 3–19 (1981)

L. Mahrt, Stably stratified atmospheric boundary layers. Annu. Rev. Fluid Mech. **46**, 23–45 (2014)

L. Mahrt, D. Vickers, Contrasting vertical structures of nocturnal boundary layers. Bound.-Layer Meteorol. **105**, 351–363 (2002)

V.K. Makin, A note on the drag of the sea surface at hurricane winds. Bound.-Layer Meteorol. **115**, 169–176 (2005)

S.P. Malinowski, K.E. Haman, M.K. Kopec, W. Kumala, H. Gerber, Small-scale turbulent mixing at stratocumulus top observed by means of high resolution airborne temperature and LWC measurements. J. Phys. Conf. Ser. **318** (2011). 072013.1–072013.7

I. Mammarella, F. Tampieri, M. Tagliazucca, M. Nardino, Turbulence perturbations in the neutrally stratified surface layer due to the interaction of a katabatic flow with a steep ridge. Environ. Fluid Mech. **5**, 227–246 (2005)

T. Mauritsen, G. Svensson, Observations of stably stratified shear-driven atmospheric turbulence at low and high Richardson numbers. J. Atmos. Sci. **64**, 645–655 (2007)

D.V. Mironov, E. Fedorovich, On the limiting effect of the Earth's rotation on the depth of a stably stratified boundary layer. Q. J. R. Meteorol. Soc. **136**, 1473–1480 (2010)

A.S. Monin, A.M. Yaglom, *Statistical Fluid Mechanics*, vol. I (MIT, Cambridge, 1971), 769 pp

A.S. Monin, A.M. Yaglom, *Statistical Fluid Mechanics*, vol. II (MIT, Cambridge, 1975), 874 pp

D.F. Nadeau, E.R. Pardyjak, C.W. Higgins, H.J.S. Fernando, M. Parlange, A simple model for the afternoon and early evening decay of convective turbulence over different land surfaces. Bound.-Layer Meteorol. **141**, 301–324 (2011)

F.T.M. Nieuwstadt, The turbulent structure of the stable, nocturnal boundary layer. J. Atmos. Sci. **41**, 2202–2216 (1984)

F.T.M. Nieuwstadt, A model for the stationary, stable boundary layer, in *Turbulence and Diffusion in Stable Environments*, ed. by J.C.R. Hunt (Clarendon, Oxford, 1985), pp. 149–220

F.T.M. Nieuwstadt, R.A. Brost, The decay of convective turbulence. J. Atmos. Sci. **43**, 532–546 (1986)

V. Nikora, Origin of the −1 spectral law in wall-bounded turbulence. Phys. Rev. Lett. **83**, 734–736 (1999)
A.M. Obukhov, Turbulence in thermally inhomogeneous atmosphere. Trudy In-ta Theoret. Geofiz. AN SSSR **1**, 95–115 (1946). In Russian
G.S. Poulos, W. Blumen, D.C. Fritts, J.K. Lundquist, J. Sun, S.P. Burns, C. Nappo, R.M. Banta, R.K. Newsom, J. Cuxart, E. Terradellas, B.B. Balsley, M. Jensen, CASES-99: a comprehensive investigation of the stable nocturnal boundary layer. Bull. Am. Meteorol. Soc. **83**, 555–581 (2002)
M.D. Powell, P.J. Vickery, T.A. Reinhold, Reduced drag coefficient for high wind speeds in tropical cyclones. Nature **422**, 279–283 (2003)
E. Sahlee, A.S. Smedman, A. Rutgersson, U. Högström, Spectra of co2 and water vapour in the marine atmospheric surface layer. Bound.-Layer Meteorol. **126**, 279–296 (2008)
M. Sastre, C. Yagüe, C. Román-Cascón, G. Maqueda, Atmospheric boundary-layer evening transitions: a comparison between two different experimental sites. Bound.-Layer Meteorol. **157**, 375–399 (2015)
A.S. Smedman, U. Högström, H. Bergstrom, The turbulence regime of a very stable marine airflow with quasi-frictional decoupling. J. Geophys. Res. **102**, 21049–21059 (1997)
A.S. Smedman, U. Högström, J.C.R. Hunt, E. Sahlee, Heat/mass transfer in the slightly unstable atmospheric surface layer. Q. J. R. Meteorol. Soc. **133**, 37–51 (2007)
Z. Sorbjan, Structure of the stably-stratified boundary layer during the SESAME-1979 experiment. Bound.-Layer Meteorol. **44**, 255–266 (1988)
Z. Sorbjan, Evaluation of local similarity functions in the convective boundary layer. Bound.-Layer Meteorol. **30**, 1565–1583 (1991)
Z. Sorbjan, Decay of convective turbulence revisited. Bound.-Layer Meteorol. **82**, 501–515 (1997)
Z. Sorbjan, Gradient-based scales and similarity laws in the stable boundary layer. Q. J. R. Meteorol. Soc. **136**, 1243–1254 (2010)
Z. Sorbjan, A.A. Grachev, An evaluation of the fluxgradient relationship in the stable boundary layer. Bound.-Layer Meteorol. **135**, 385–405 (2010)
M.A. Strunin, T. Hiyama, J. Asanuma, T. Ohata, Aircraft observations of the development of thermal internal boundary layers and scaling of the convective boundary layer over non-homogeneous land surfaces. Bound.-Layer Meteorol. **111**, 491–522 (2004)
J. Sun, C.J. Nappo, L. Mahrt, D. Belusic, B. Grisogono, D.R. Stauffer, M. Pulido, C. Staquet, Q. Jiang, A. Pouquet et al., Review of wave-turbulence interactions in the stable atmospheric boundary layer. Rev. Geophys. **53**, 956–993 (2015)
F. Tampieri, A. Maurizi, A. Viola, An investigation on temperature variance scaling in the atmospheric surface layer. Bound.-Layer Meteorol. **132**, 31–42 (2009)
F. Tampieri, C. Yagüe, S. Viana, The vertical structure of second-order turbulence moments in the stable boundary layer from SABLES98 observations. Bound.-Layer Meteorol. **157**, 45–59 (2015)
I. Troen, L. Mahrt, A simple model of the atmospheric boundary layer: sensitivity to surface evaporation. Bound.-Layer Meteorol. **39**, 129–148 (1986)
W.S. Uijttewaal, R.V.A. Oliemans, Particle dispersion and deposition in direct numerical and large eddy simulations of vertical pipe flows. Phys. Fluids **8**, 2590–2604 (1996)
A.P. van Ulden, A.A.M. Holtslag, Estimation of atmospheric boundary layer parameters for diffusion applications. J. Clim. Appl. Meteorol. **24**, 1196–1207 (1985)
J.C. Wyngaard, O.R. Coté, The budgets of turbulent kinetic energy and temperature variance in the atmospheric surface layer. J. Atmos. Sci. **28**, 190–201 (1971)
J.C. Wyngaard, O.R. Coté, Cospectral similarity in the atmospheric surface layer. Q. J. R. Meteorol. Soc. **98**, 590–603 (1972)
J.C. Wyngaard, O.R. Coté, Y. Izumi, Local free convection, similarity, and the budgets of shear stress and heat flux. J. Atmos. Sci. **28**, 1171–1182 (1971)
A.M. Yaglom, Fluctuation spectra and variances in convective turbulent boundary layers: a reevaluation of old models. Phys. Fluids **6**, 962–972 (1994)

C. Yagüe, S. Viana, G. Maqueda, J.M. Redondo, Influence of stability on the flux-profile relationships for wind speed, ϕ_m, and temperature, ϕ_h, for the stable atmospheric boundary layer. Nonlinear Process. Geophys. **13**, 185–203 (2006)
S.S. Zilitinkevich, Comments on "A model for the dynamics of the inversion above a convective boundary layer". J. Atmos. Sci. **32**, 991–992 (1975)
S.S. Zilitinkevich, Third-order transport due to internal waves and non-local turbulence in the stably stratified surface layer. Q. J. R. Meteorol. Soc. **128**, 913–925 (2002)
S.S. Zilitinkevich, The height of the atmospheric planetary boundary layer: state of the art and new development, in *NATO Science for Peace and Security Series C: Environmental Security* (Springer, Berlin, 2012), pp. 147–161
S.S. Zilitinkevich, A. Baklanov, Calculation of the height of the stable boundary layer in practical applications. Bound.-Layer Meteorol. **105**, 389–409 (2002)
S.S. Zilitinkevich, I.N. Esau, On integral measures of the neutral barotropic planetary boundary layer. Bound.-Layer Meteorol. **104**, 371–379 (2002)
S.S. Zilitinkevich, I.N. Esau, Resistance and heat-transfer laws for stable and neutral planetary boundary layers: old theory advanced and re-evaluated. Q. J. R. Meteor. Soc. **131**, 1863–1892 (2005)
S.S. Zilitinkevich, I.N. Esau, Similarity theory and calculation of turbulent fluxes at the surface for the stably stratified atmospheric boundary layer. Bound.-Layer Meteorol. **125**, 193–205 (2007)
S.S. Zilitinkevich, D.V. Mironov, A multi-limit formulation for the equilibrium depth of a stably stratified boundary layer. Bound.-Layer Meteorol. **81**, 325–351 (1996)
S.S. Zilitinkevich, A.A. Grachev, J.C.R. Hunt, Surface frictional processes and non-local heat/mass transfer in the shear-free convective boundary layer, in *Buoyant Convection in Geophysical Flows*, ed. by E.J. Plate (Kluwer, Dordrecht, 1998)
S.S. Zilitinkevich, J.C.R. Hunt, I.N. Esau, A.A. Grachev, D.P. Lalas, E. Akylas, M. Tombrou, C.W. Fairall, H.J.S. Fernando, A. Baklanov, S.M. Joffre, The influence of large convective eddies on the surface-layer turbulence. Q. J. R. Meteorol. Soc. **132**, 1423–1456 (2006)
S.S. Zilitinkevich, I.N. Esau, A. Baklanov, Further comments on the equilibrium height of neutral and stable planetary boundary layers. Q. J. R. Meteorol. Soc. **133**, 265–271 (2007)
S.S. Zilitinkevich, T. Elperin, N. Kleeorin, I. Rogachevskii, I.N. Esau, T. Mauritsen, M.W. Miles, Turbulence energetics in stably stratified geophysical flows: strong and weak mixing regimes. Q. J. R. Meteorol. Soc. **134**, 793–799 (2008)
S.S. Zilitinkevich, T. Elperin, N. Kleeorin, I. Rogachevskii, I.N. Esau, A hierarchy of energy- and flux-budget (EFB) turbulence closure models for stably-stratified geophysical flows. Bound.-Layer Meteorol. **146**, 341–373 (2013)

Chapter 4
Horizontal Heterogeneities

Abstract In this Chapter we address a number of issues that contradict the fundamental requirements of the basic paradigm presented in Chap. 3. In this way we tackle the question, how to cope with the real world features. Two broad lines of approach are presented: the explicit treatment of the space-dependent problem, and the averaging over the typical variation scale. Changes in the surface features and topographic effects are examples for the first approach. Flow into a canopy (both a forest and a city) and parameterizations of sub-grid scale (unresolved) heterogeneities are examples for the second approach.

4.1 Explicit Treatment vs. Parameterization

The paradigm of Chap. 3 is difficult to be fully realized in practice (note the effort put in order to individuate 'ideal' conditions in the planning of the field experiments).

As discussed in Sect. 1.3.2, the problem will be tackled in the following Sections for a number of key examples by means of an explicit treatment of the heterogeneity effects (large scale heterogeneities, like a sharp change of surface roughness), or by parameterization of the overall effects (small scale heterogeneities, like random distribution of roughness elements). Finally, the urban boundary layer is treated as an exemplar situation in which the two approaches coexist.

We will face in this Chapter also the low-wind conditions, which are an important example of fault of the similarity theory and of small scale (from our point of view) unsteadiness. Large scale unsteadiness, as produced by the diurnal cycle, has been investigated in the previous Chapter.

4.1.1 A Criterion for Horizontal Homogeneity

Horizontal homogeneity is evidently an abstraction. It may be useful to examine under which conditions the equations for the TKE and the temperature variance in the horizontally homogeneous form, like (3.10), can be used: to this end, we can consider the steady, 2D equation for TKE, neglecting the third-order term

F. Tampieri, *Turbulence and Dispersion in the Planetary Boundary Layer*,
Physics of Earth and Space Environments, DOI 10.1007/978-3-319-43604-3_4

divergence:

$$U\frac{\partial E_{\mathrm{k}}}{\partial x}+W\frac{\partial E_{\mathrm{k}}}{\partial z}+\left(\langle u^2\rangle-\langle w^2\rangle\right)\frac{\partial U}{\partial x}+\langle uw\rangle\frac{\partial W}{\partial x}=-\langle uw\rangle\frac{\partial U}{\partial z}\left(1-R_{\mathrm{f}}\right)-\varepsilon\;, \tag{4.1}$$

where we have put on the LHS the terms related to the horizontal derivatives and on the RHS the terms present in the homogeneous case.

Clearly, to neglect the LHS, it is necessary that the horizontal scales of variation of the mean velocity and of the TKE, $|U(\partial U/\partial x)^{-1}|$ and $|E_{\mathrm{k}}(\partial E_{\mathrm{k}}/\partial x)^{-1}|$, turn out to be larger than the vertical scale $|U(\partial U/\partial z)^{-1}|$, given that the variances of the velocity components and the TKE are of the same order, and that the flux Richardson number is far from the critical value.

The horizontal derivatives are intuitively related to the presence of changes of the surface features. But even in nearly horizontally homogeneous conditions, the horizontal scale related to the mean velocity becomes small if the wind velocity approaches zero, allowing a great sensitivity to horizontal heterogeneities. Also, the vertical production term tends to zero in strong stable conditions, for near critical conditions, making the RHS small and thus enhancing the importance of the LHS.

As far the temperature variance is considered, the equation under the same conditions reads:

$$U\frac{\partial\langle\theta^2\rangle}{\partial x}+W\frac{\partial\langle\theta^2\rangle}{\partial z}+2\langle u\theta\rangle\frac{\partial\theta_{\mathrm{f}}}{\partial x}=-2\langle w\theta\rangle\frac{\partial\theta_{\mathrm{f}}}{\partial z}-\varepsilon_\theta\;. \tag{4.2}$$

As in the TKE case, the horizontal heterogeneity can be neglected if the horizontal scales of variation are larger than the vertical scale for θ_{f}. From this equation, it is evident that even if the vertical heat flux is negligible, horizontal heterogeneities can balance dissipation, leading to finite temperature variance in quasi-neutral conditions, as shown in Sect. 3.4.3.4.

4.2 Internal Boundary Layers

Here we discuss in short the internal boundary layers (IBL) that develop when there is a change of surface conditions: in particular, the effects of sudden changes of the roughness length and of the surface heat flux.

Internal boundary layers are found in the CBL near the surface, in the regions of convergence at the foot of the updrafts, under condition of very low average horizontal wind (shear-free boundary layers): these will not be treated in this book.

4.2.1 Roughness Length Changes

We shall use a schematic, two-dimensional approach. Let consider a flat surface with a change in the value of the roughness length from $z_0 = z_{01}$ for $x < 0$ to $z_0 = z_{02}$ for $x \geq 0$ in the direction of the mean wind. In neutral conditions, for $x < 0$ (upwind the roughness change):

$$U_{\mathrm{U}}(z) = \frac{u_{*1}}{\kappa} \ln(z/z_{01}), x < 0 \; . \tag{4.3}$$

Downwind of the change, we assume that the boundary layer is in equilibrium with the new roughness value, and a logarithmic profile for the mean wind with a different friction velocity is found. The two profiles match at the height h_{i}, the depth of the internal boundary layer. For $z > h_{\mathrm{i}}$ the upstream profile is found. Thus

$$U(x, z) = \frac{u_{*2}}{\kappa} \ln(z/z_{02}), x > 0, z < h_{\mathrm{i}} \; , \tag{4.4}$$

where the x dependence comes from the IBL depth via the local friction velocity.

Matching (4.3) and (4.4) at $z = h_{\mathrm{i}}$, it results:

$$\frac{u_{*1}}{u_{*2}} = \frac{\ln(h_{\mathrm{i}}/z_{02})}{\ln(h_{\mathrm{i}}/z_{01})} \; . \tag{4.5}$$

An estimate for h_{i} may be obtained as follows. For a smooth-to-rough transition, we assume that the IBL growth is driven by a diffusive process (after Miyake 1965), whose effect is parameterized through the variance of the vertical velocity fluctuations (see Savelyev and Taylor 2005, for a summary) in the IBL. Thus, because the vertical velocity variance scales on the square of the friction velocity:

$$\frac{\mathrm{d}h_{\mathrm{i}}}{\mathrm{d}x} = \frac{\alpha u_{*2}}{U(x, h_{\mathrm{i}})} = \alpha\kappa \left[\ln\left(\frac{h_{\mathrm{i}}}{z_{02}}\right) \right]^{-1} \; , \tag{4.6}$$

where the numerical value of $\alpha \sim 1$ relates to the efficiency of mixing. In the case of rough-to-smooth transition, the turbulence (and the mixing) is less intense in the lower than in the upper layer, according to (4.5), and the IBL growth is likely to be dominated by the upwind features of the turbulence, measured by u_{*1} and z_{01}. Thus, in the practical application of the formula, the mixing at the top can be assumed to be driven by the turbulence related to the higher value of the roughness, which is the same argument that lead to the Wood (1982) formula. The integration of (4.6) gives:

$$\frac{h_{\mathrm{i}}}{z_{0\mathrm{e}}} \left(\ln \frac{h_{\mathrm{i}}}{z_{0\mathrm{e}}} - 1 \right) + 1 = \alpha\kappa \left(\frac{x}{z_{0\mathrm{e}}} - 1 \right) \; , \tag{4.7}$$

where instead of z_{02} we use $z_{0\mathrm{e}} = \max[z_{01}, z_{02}]$.

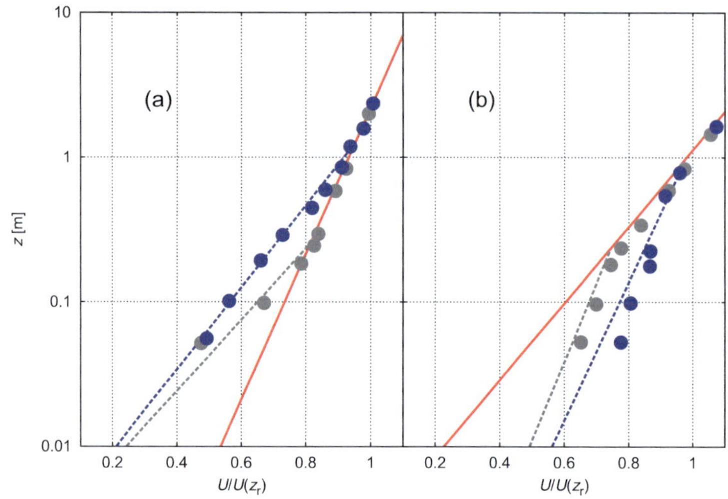

Fig. 4.1 Wind profiles for roughness changes. Data from Bradley (1968), reported by Garratt (1990): *gray* and *blue symbols* refer to the two different positions downwind the roughness change (see below). The two roughness lengths are $z_{01} = 2 \times 10^{-5}$ m, $z_{02} = 2.5 \times 10^{-3}$ m. The reference velocity from the upwind profile (*red line*) is measured at $z_r = 2.2$ m in (**a**) and at $z_r = 1.125$ m in (**b**). (**a**) Smooth-to-rough transition; *gray line*: downwind profile at $x = 2.32$ m, from (4.4), (4.5) and (4.7) with $\alpha = 1$; *blue line*: the same at $x = 16.42$ m. (**b**) Rough-to-smooth transition; *gray line*: downwind profile at $x = 2.1$ m; *blue line*: the same at $x = 12.2$ m

This simple model is compared with field observations in Figs. 4.1 and 4.2. The observed decay of the friction velocity downwind of the change is faster than that predicted by the simple model, while the asymptotic value is correctly reproduced; also h_i is fairly described.

The changes of surface characteristics are a common feature of the real world. Thus we expect that a real wind profile is made by the superposition of profiles characterized by different values of the roughness length and the friction velocity, in neutral conditions. (The situation is more complex in presence of heat fluxes.)

A practical consequence of the IBL formation is that the wind measured at a given height (on a mast) is affected by the features of the surrounding terrain, and as the height increases, the area influencing the measure increases. In this way, the profile features (namely, the roughness length and the momentum flux consistent with the observed mean velocity) are in general changing with height: this is a typical effect of the heterogeneity of the surface.

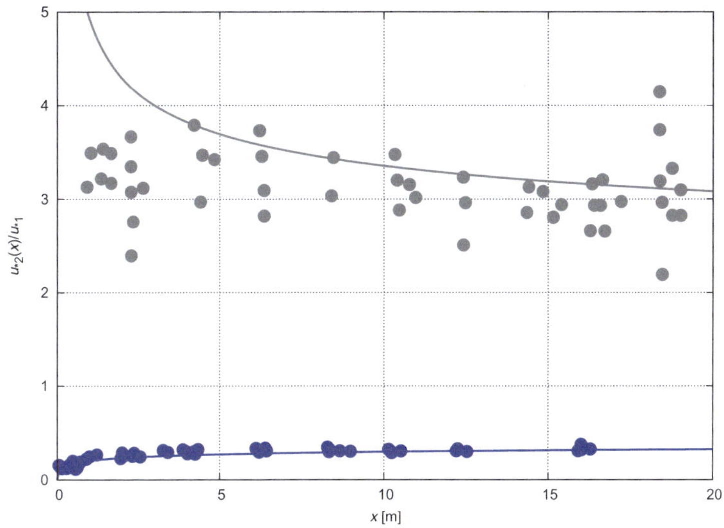

Fig. 4.2 Ratio of the downwind friction velocity to the upwind one, for roughness changes, as function of the distance downwind the change. *Gray symbols and line*: smooth-to-rough transition; *blue symbols and line*: rough-to-smooth transition. Data as in Fig. 4.1; lines from (4.5) and (4.7)

4.2.2 The Thermal IBL at the Sea-Land Transition

The sea-land transition is a second example of IBL: assume a neutral or stable flow from the (relatively cold) sea, producing unstable conditions over the (relatively warm) land. This situation has been widely studied and simulated, also in the frame of dispersion modelling. The reversed situation can also occur, SBL over the sea downwind a neutral or unstable PBL over land (Rogers et al. 1995).

The results obtained in Sect. 3.6.2 are applied here by substituting $x/\overline{U}$ to t in (3.91) and in the equations that derive from it. Here the velocity $\overline{U}$ has a meaning similar to the mean temperature averaged in the CBL, i.e. represents the advection of air as a bulk process. If we characterize the land surface with the heat flux $\langle \theta w \rangle_0$ (assumed constant with the downwind distance x from the shoreline), and we refer to the model discussed in Sect. 3.6.2.2, the depth of the internal boundary layer is a function of the inland distance according to:

$$h_{\rm i}(x) = \left[\frac{2\langle \theta w \rangle_0 (2A + 1)}{\Gamma \overline{U}} \right]^{1/2} x^{1/2} , \tag{4.8}$$

for a case dominated by the surface heat flux, and:

$$h_{\rm i}(x) = \left[\frac{6\, B u_*^3 \theta_{00}}{g \Gamma \overline{U}} \right]^{1/3} x^{1/3} , \tag{4.9}$$

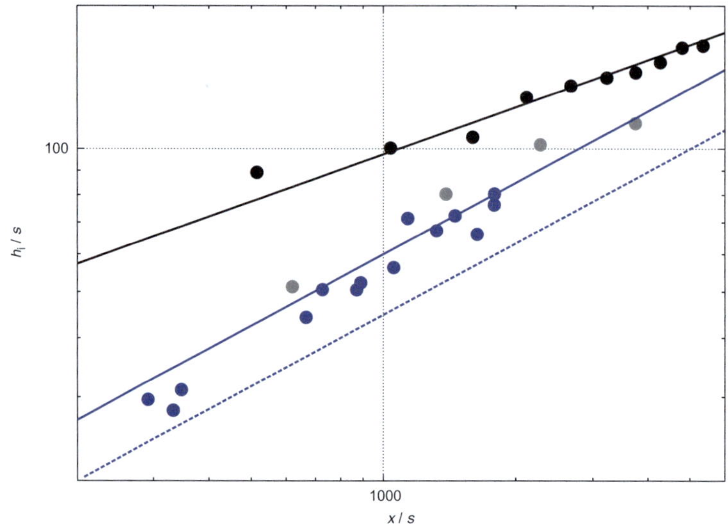

Fig. 4.3 IBL depth at the equilibrium, after a change in surface heat flux, from different field experiments (Villani et al. 2005). The *blue symbols* refer to three cases with Obukhov length $L \simeq -45$ m, the *black* ones to a case with $L \simeq -76$ m, the *gray* ones to an intermediate situation. The *blue continuous line* corresponds to (4.8) with $A = 0.4$, the *blue dashed line* to the encroachment model, i.e. (4.8) with $A = 0$; the *black line* to (4.9) with $B = 10$. It is evident that the entrainment at the top enhances the growth, and the encroachment model underestimates the IBL depth. It is evident also that the less unstable case shows a growth better described by the $1/3$ power law

for a case dominated by the friction at the surface, with the same meaning of the symbols as in Sect. 3.6.2. Thus the depth of this two-dimensional IBL is expected to grow as a power law of the distance from the shoreline, with exponent between $1/2$ and $1/3$.

In Fig. 4.3 some literature cases are presented, see Gryning and Batchvarova (1990, Table 1) and Villani et al. (2005, Fig. 4). The downwind distance and the IBL depth are normalized with the scale appropriate of the encroachment model:

$$s = \frac{\langle \theta w \rangle_0}{\Gamma \overline{U}} , \tag{4.10}$$

so that for a case without entrainment and dominated by the heat flux the nondimensional depth is $\sqrt{2}$ times the square root of the nondimensional distance.

Note that the internal boundary layer is assumed to be in local equilibrium with the surface heat flux and this means that this description holds when the unstable boundary layer over land is fully established. More complex models can be found in the literature: for a summary of the analytical models see Gryning and Batchvarova (1990). Numerical simulations allow to explicitly describe the onset of the sea breeze, before the equilibrium is reached (see, for instance Antonelli and Rotunno 2007).

4.3 The Boundary Layer Over Hills and Valleys

In this Section we discuss the basic effects of the small scale topographic features (hills and valleys of scale height H_t much smaller than the PBL depth, and of gentle slope: $L_t \gg H_t$, being L_t a streamwise scale length of the topography). With these conditions we refer to linearized perturbations to the flow, although some results can be applied to a wider range of conditions.

The interaction of a boundary layer with a topographic feature has a twofold aspect:

- a local modification of the PBL, which can be understood in terms of presence or absence of local equilibrium;
- a large-scale perturbation, due both to the turbulent wake and to the stratification effects.

The study of flow over hills is relevant for instance for the planning of wind farms, although since long time the windmills were placed on the slopes of the hills, and not in valleys: the landscape seen by Don Quijote (De Cervantes 1605). A large number of studies are found in the literature: Berg et al. (2011) give a review of field and wind tunnel experiments; in particular, the extensive laboratory experiment of flow over hills and valleys by Khurshudyan et al. (1981, 1990) and Snyder et al. (1991) can be cited.

4.3.1 The Linearized Equations

Equation (2.68), neglecting the Coriolis term, (2.64) and (2.65) in two dimensions (x and z) are written for the small quantities ($\Delta U, W$), ΔP, Θ, being the (upstream) basic state $(U_U, 0)$ and $\theta_{00} + \theta_0$ (the basic state for pressure being zero, remember the hydrostatic balance). The molecular diffusion terms are neglected.

The horizontal momentum equation reads:

$$U_U \frac{\partial \Delta U}{\partial x} + W \frac{dU_U}{dz} = -\frac{1}{\rho_{00}} \frac{\partial \Delta P}{\partial x} - \frac{\partial \langle u^2 \rangle}{\partial x} + \frac{\partial \Delta \tau}{\partial z} , \tag{4.11}$$

where $\tau \equiv -\langle uw \rangle = u_*^2 + \Delta\tau$ and u_* is constant; the vertical momentum equation reads:

$$U_U \frac{\partial W}{\partial x} = -\frac{1}{\rho_{00}} \frac{\partial \Delta P}{\partial z} + \frac{\partial \Delta \tau}{\partial x} - \frac{\partial \langle w^2 \rangle}{\partial z} + \frac{g}{\theta_{00}} \Theta ; \tag{4.12}$$

the continuity equation reads:

$$\frac{\partial \Delta U}{\partial x} + \frac{\partial W}{\partial z} = 0 \; ; \tag{4.13}$$

and the temperature equation reads:

$$U_{\mathrm{U}} \frac{\partial \Theta}{\partial x} + W \frac{\mathrm{d}\theta_0}{\mathrm{d}z} = -\frac{\partial \langle u\theta \rangle}{\partial x} - \frac{\partial \langle w\theta \rangle}{\partial z} \; . \tag{4.14}$$

The perturbation variables are of the order of the ratio $H_{\mathrm{t}}/L_{\mathrm{t}} \ll 1$. Because the second-order moments of velocity components are of the same order: $\langle u^2 \rangle \sim \langle uw \rangle$, it results that the horizontal gradient of $\langle u^2 \rangle$ is in magnitude smaller than the vertical gradient of $\langle uw \rangle$, so only the vertical derivative can be kept in the RHS of (4.11). The same reasoning leads to drop the horizontal gradient of the horizontal heat flux in (4.14).

4.3.2 *The Inner and Outer Layer Concept in the Neutral Flow*

Let consider the wind profile upstream of the obstacle $U_{\mathrm{U}}(z) = u_*/\kappa \ln(z/z_0)$. Assume that the roughness length is constant in space. The travel time of a fluid parcel at height z over the obstacle is $\sim L_{\mathrm{t}}/U_{\mathrm{U}}(z)$ while the Eulerian integral time scale is $\kappa z/u_*$. The integral time scale is a measure of the time characteristic for a diffusive process: we can imagine that near the surface the flow is in equilibrium because the flux-gradient relationship holds. The time for the diffusion process to reach the equilibrium at z is thus $\alpha\kappa z/u_*$ where $\alpha \geq 1$ The former time scale decreases and the latter increases with z, identifying a level at which are equal. Thus we define the depth l_{t} of the inner layer:

$$\frac{l_{\mathrm{t}}}{L_{\mathrm{t}}} \ln \frac{l_{\mathrm{t}}}{z_0} = \alpha^{-1} \; , \tag{4.15}$$

such that for $z < l_{\mathrm{t}}$ the turbulence is expected to be in local equilibrium, while for $z > l_{\mathrm{t}}$ the eddies (at least, the larger eddies) are advected too fast to reach equilibrium conditions with the perturbation produced by the obstacle. Note that the internal boundary layer depth formula (4.7) is based on the same ideas as (4.15).

On the other hand, following the analysis by Jackson and Hunt (1975), for a neutral flow over a gentle slope topography, we can consider the balance between the streamwise advection and the vertical divergence of the momentum flux, from the linearized momentum Eq. (4.11). The advection term reads:

$$U \frac{\partial \Delta U}{\partial x} \approx U_{\mathrm{U}} \frac{\Delta U}{L_{\mathrm{t}}} \; , \tag{4.16}$$

while the momentum flux divergence term reads:

$$\frac{\partial \Delta\tau}{\partial z} \approx \frac{\Delta\tau}{z} \,, \tag{4.17}$$

The height at which the two terms are equivalent is a further estimate of the depth of the inner layer l_t.

The flux-gradient relationship for the logarithmic profile states that the eddy diffusion coefficient is $\kappa u_* z$, as discussed in Sect. 3.4.2.3. Assuming that in the inner layer, over the obstacle, the eddy diffusion coefficient is $\kappa \tau^{1/2} z$, the flux-gradient relationship reads:

$$\tau = \kappa \tau^{1/2} z \left[\frac{\mathrm{d}U_\mathrm{U}}{\mathrm{d}z} + \frac{\partial \Delta U}{\partial z} \right] , \tag{4.18}$$

which becomes

$$\tau = \kappa^2 z^2 \left[\frac{\mathrm{d}U_\mathrm{U}}{\mathrm{d}z} + \frac{\partial \Delta U}{\partial z} \right]^2 , \tag{4.19}$$

consistent with (2.84). From (4.19), neglecting the square of the perturbation derivative, it results:

$$\Delta\tau = 2\kappa u_* z \frac{\partial \Delta U}{\partial z} \,. \tag{4.20}$$

Using this expression in (4.17) the following estimate for the momentum flux divergence arises:

$$\frac{\partial \Delta\tau}{\partial z} \approx 2\kappa u_* \frac{\partial \Delta U}{\partial z} \approx 2\kappa u_* \frac{\Delta U}{z} ;. \tag{4.21}$$

Thus the depth of the inner layer corresponds to the height at which (4.16) and (4.21) are equal:

$$\frac{l_\mathrm{t}}{L_\mathrm{t}} \ln \frac{l_\mathrm{t}}{z_0} = 2\kappa^2 \,. \tag{4.22}$$

This estimate of the inner layer depth was given by Jackson and Hunt (1975).

If the perturbed profile is assumed to be logarithmic $\Delta U = \alpha \ln(z/z_0)$ (Jensen et al. 1984), from (4.20) it results $\Delta\tau = 2\kappa u_* \Delta U / \ln(z/z_0)$ and the estimate of the inner layer depth turns out to be:

$$\frac{l_\mathrm{t}}{L_\mathrm{t}} \left[\ln \frac{l_\mathrm{t}}{z_0} \right]^2 = 2\kappa^2 \,. \tag{4.23}$$

This estimate is quite consistent with the data, and widely used (Taylor et al. 1987).

It is important to notice that different approaches lead to different values of the depth of the inner layer, but the concept is unique. Below l_t (in the inner layer) the wind profile is approximated by a logarithmic one, characterized by the local value of the friction velocity; above such level (in the outer layer), the perturbation to the flow is essentially non-turbulent (the cross-section of the flow is reduced in presence of a hill, and the flow accelerates; viceversa for a valley).

If in the inner layer the wind profile is logarithmic and the roughness length is the same upwind and over the hill, the relative increment of the mean velocity (the speed-up over the hill) is constant with height, and reads:

$$\Delta S \equiv \frac{U(x,z) - U_U(z)}{U_U(z)} = \frac{\tau^{1/2}}{u_*} - 1 , \tag{4.24}$$

where $\tau(x, 0)$ is the absolute value of the momentum flux at the surface in the position x along the cross-section of the hill. Note that (4.24) is a relation between the changing shear stress and the speed-up.

The simple results outlined above are summarized and compared with data in Fig. 4.4 which is based on the observations made during the Askervein Hill experiment (Zeman and Jensen 1987). The observations show the upstream logarithmic

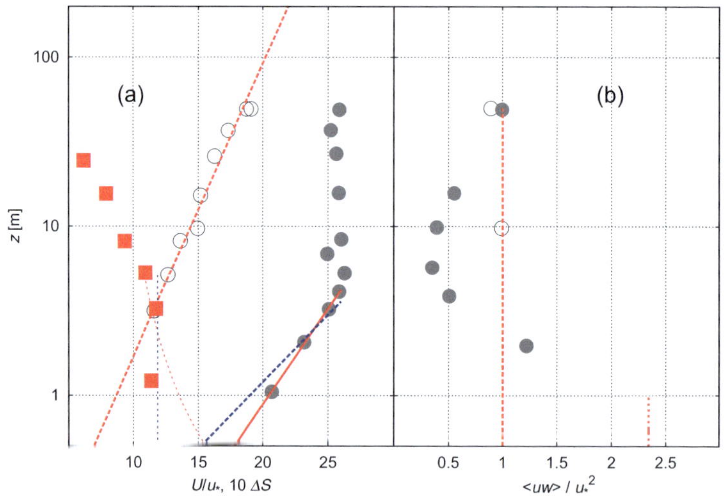

Fig. 4.4 Mean wind (**a**) and momentum flux (**b**) upstream (*open circles*) and at the top (*full circles*) of the Askervein Hill (redrawn from Zeman and Jensen 1987, Figs. 3 and 4). In (**a**) the speed up (redrawn from Moreira et al. 2012, Fig. 8) is reported as *red squares*. In (**a**) the *red dashed line* is the fit to the upstream wind data, the *continuous red line* the fit to the inner layer data at the top, the *blue dashed line* the fit forcing the roughness length to be the same as upstream. The *red and blue thin dotted lines* are the corresponding speed-up values. The *red dashed line* in (**b**) highlights the surface layer upstream, the *dotted line* highlights the increase in momentum flux at the surface, as derived from the velocity fits. Note that with $z_0 = 0.03$ m and $L_t = 250$ m, (4.22) gives $l_t \approx 13$ m while (4.23) gives $l_t \approx 3.5$ m

wind profile and the almost logarithmic wind profile at the hill top, from the surface up to $l_t \simeq 6\,\mathrm{m}$. At about the same height the momentum flux shows a minimum, and increases going towards the surface.

4.3.3 *The Outer Layer and the Stratification Effects*

To give a description of the basic behaviour of the flow in the outer layer (region), we consider the 2D equations for momentum and temperature (Sect. 4.3.1) and neglect the turbulent fluxes (remember that the perturbations are inviscid: see Sect. 4.3.2).

4.3.3.1 The Linearized Equation for the Vertical Velocity

By differentiating (4.11) by z and (4.12) by x, and substituting with suitable quantities from (4.13) and (4.14), the equation for W reads:

$$\frac{\partial^2 W}{\partial x^2} + \frac{\partial^2 W}{\partial z^2} + \left(\frac{N^2}{U_U^2} - \frac{1}{U_U} \frac{d^2 U_U}{dz^2} \right) W = 0\,, \tag{4.25}$$

whose Fourier transform is:

$$\frac{\partial^2 \widehat{W}}{\partial z^2} + \left(\frac{N^2}{U_U^2} - \frac{1}{U_U} \frac{d^2 U_U}{dz^2} - k^2 \right) \widehat{W} = 0\,. \tag{4.26}$$

This equation describes the internal stationary waves (Lighthill 1978). About atmospheric waves generated by the hills: see Gill (1982, p. 268).

The Scorer parameter:

$$l_S^{-2} = \frac{N^2}{U_U^2} - \frac{d^2 U_U / dz^2}{U_U} \tag{4.27}$$

accounts for the stability of the basic state and the curvature effect of the upstream wind profile.

For the SBL, as far the log-linear profile is valid, the curvature term is negligible as $z/\Lambda > 1$, thus (4.27) is affected mainly by the upstream stability.

4.3.3.2 A Simple Solution for the Outer Layer

We shall neglect the vertical variations of the upstream profile: U_U constant with height. Thus (4.26) becomes:

$$\frac{\partial^2 \widehat{W}}{\partial z^2} + \left(\frac{N^2}{U_U^2} - k^2\right)\widehat{W} = 0 \ , \tag{4.28}$$

and its solution reads:

$$\widehat{W} = A\exp(-\imath mz) + B\exp(\imath mz) \ . \tag{4.29}$$

The lower boundary condition is given by:

$$\frac{W(x, z = H_t f)}{U(x, z = H_t f)} = H_t\frac{df}{dx} \ , \tag{4.30}$$

which states that the flow follows the shape of the topography, described by $H_t f(x)$. This condition is linearized as follows:

$$\frac{W(x, 0)}{U_U} = H_t\frac{df}{dx} \ , \tag{4.31}$$

and determines the form of $(A + B)$. The application of the nonlinear condition is discussed by Lilly and Klemp (1979).

The upper boundary condition is twofold: the perturbation decays with height, or the energy of the waves propagates upwards. Thus, if m is imaginary, $A = 0$ gives a solution decaying with height. The dispersion relation of the gravity waves in the 2D case reads (Lighthill 1978, p. 296):

$$\omega^2 = N^2\frac{k^2}{k^2 + m^2} \ . \tag{4.32}$$

The waves must be stationary with respect to the ground, i.e. the horizontal component of their phase velocity must be proportional to minus the wind velocity: $\omega/k \propto -U_U$; to satisfy the radiation condition (upwards transport of energy) the vertical component of the group velocity must be positive: $\partial\omega/\partial m \equiv -\omega m/(k^2 + m^2) > 0$ so that the radiation condition for stationary waves results $\mathrm{sgn}(m) = \mathrm{sgn}(k)$.

Finally, the solution to (4.29) results:

$$\widehat{W} = \imath k U_U H_t \hat{f} \exp(\imath mz) \ , \tag{4.33}$$

where m is given by the decaying condition:

$$m = \imath \left(k^2 - \frac{N^2}{U_U^2} \right)^{1/2} \quad \text{for } \frac{N^2}{U_U^2} < k^2 \,, \tag{4.34}$$

or the upwards radiation condition

$$m = \text{sgn}(k) \left(\frac{N^2}{U_U^2} - k^2 \right)^{1/2} \quad \text{for } \frac{N^2}{U_U^2} > k^2 \,. \tag{4.35}$$

Note that (4.34) and (4.35) suggest that the decaying perturbation occurs for wavenumbers larger than a threshold defined by the upstream wind and stability, the oscillating perturbation for smaller wavenumbers.

From the continuity equation results:

$$\imath k \widehat{\Delta U} = -\,\mathrm{d}\widehat{W} / \,\mathrm{d}z \,, \tag{4.36}$$

so the horizontal component of the perturbation velocity reads:

$$\widehat{\Delta U} = -\imath m U_U H_t \hat{f} \exp(\imath m z) \tag{4.37}$$

and, from (4.11), the expression for the pressure perturbation results:

$$\widehat{\Delta P} = -\rho_{00} U_U \widehat{\Delta U} \,. \tag{4.38}$$

To understand the main features of the outer layer solution, a Gaussian-shaped topography is considered: $f = \exp[-(x/L_t)^2]$, whose total length is about $6L_t$. Its Fourier transform is:

$$\hat{f} = \frac{L_t}{2\sqrt{\pi}} \exp[-(kL_t)^2/4] \,, \tag{4.39}$$

so that the vertical velocity results:

$$W = \imath \frac{U_U H_t}{2\sqrt{\pi} L_t} \int_{-\infty}^{\infty} \chi \exp[-\chi^2/4] \exp(\imath m z) \,\mathrm{d}\chi \,, \tag{4.40}$$

where $\chi = kL_t$ is the nondimensional wavenumber.

Similarly, the perturbation to the horizontal component of the velocity reads:

$$\Delta U = -\imath \frac{U_U H_t}{2\sqrt{\pi} L_t} \int_{-\infty}^{\infty} \mu(\chi) \exp[-\chi^2/4] \exp(\imath m z) \,\mathrm{d}\chi \,, \tag{4.41}$$

where $\mu = mL_t$.

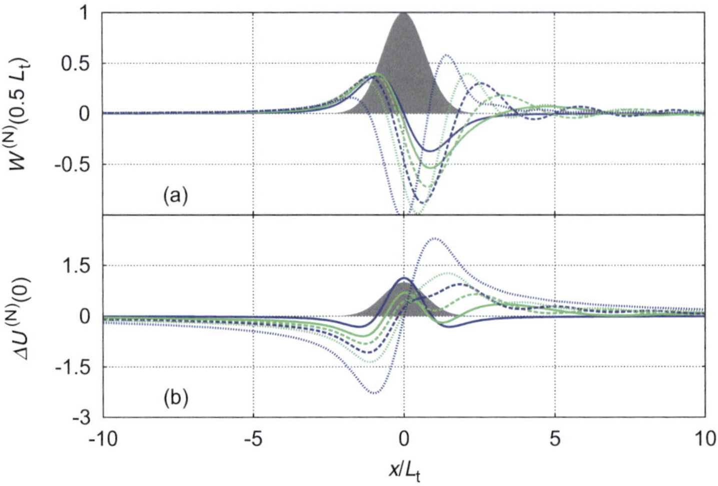

Fig. 4.5 Normalized perturbation to the vertical velocity at $z = L_t/2$ (**a**) and of the horizontal velocity at the bottom of the outer region (**b**) over a Gaussian topography $H_t \exp[-(x/L_t)^2]$ as function of x/L_t for different values of F_l^{-1}. *Blue continuous line*: $F_l^{-1} = 0$; *green continuous*: $= 1$; *green dashed* $= 1.5$, *blue dashed*: $= 2$, *green dotted*: $= 2.5$, *blue dotted*: $= 4$

The linear perturbation is proportional to the ratio H_t/L_t as expected. From (4.34) and (4.35) it results that $\mu \propto |F_l^{-2} - \chi^2|^{1/2}$, i.e. the wave perturbation occurs for nondimensional wavelengths larger than the inverse of the Froude number. For conditions approaching neutrality ($F_l \gg 1$) the vertical velocity at the ground takes the form of the derivative of the obstacle (the slope) as expected; in stable conditions, the phase moves upstream with height, according to the radiation condition; for conditions of large stability ($F_l \ll 1$) the horizontal component of the velocity increases with stability, being the coefficient of the integral proportional to $H_t N$. The largest contribution to the vertical velocity perturbations comes from the extremes of the function $\chi \exp(-\chi^2/4)$, located at $\chi = \pm\sqrt{2}$, corresponding to $k = \pm\sqrt{2}/L_t$. In other words, we can expect that for undulatory perturbation the waves with larger amplitude have a wavelength $\lambda \equiv 2\pi/k = \sqrt{2}\pi L \approx 5L_t$ that approximately matches the size of the hill.

In Fig. 4.5 some solutions are reported, in terms of perturbation velocities normalized by dividing by $U_U H_t/L_t$.

4.3.4 A Discussion About the Inner Layer

The linearized solution in the inner layer is not straightforward, and has been discussed by many authors (see Nanni and Tampieri 1985; Hunt et al. 1988a,b). An hint about the flow behaviour can be obtained neglecting the stability and considering the pressure as given by the outer layer solution (independent on height in the inner layer). The rationale is that the inner layer is shallow.

The horizontal momentum Eq. (4.11) is simplified as follows:

$$U_U \frac{\partial \Delta u}{\partial x} = -\frac{1}{\rho_{00}} \frac{d\Delta P}{dx} + \frac{\partial \Delta \tau}{\partial z} , \tag{4.42}$$

where we use Δu to refer to the inner layer velocity perturbation, in order to avoid confusion with the outer layer perturbations, that are indicated with capital letters; ΔP is the outer layer pressure perturbation at $z = 0$. Note also that we use z for the height both in the inner and in the outer layer: but $z = 0$ for the outer layer means broadly the top of the inner layer, $\approx l_t$.

Using the expressions (4.17) and (4.21), it results:

$$\frac{u_*}{\kappa} \ln\left(\frac{z}{z_0}\right) \frac{\partial \Delta u}{\partial x} = -\frac{1}{\rho_{00}} \frac{d\Delta P}{dx} + \frac{2\kappa u_*}{z} \Delta u , \tag{4.43}$$

which gives, at the surface $z = z_0$, the simple relation for the surface shear stress perturbation:

$$\Delta \tau_0 = \frac{z_0}{\rho_{00}} \frac{d\Delta P}{dx} , \tag{4.44}$$

that is, the shear stress at the surface is in phase with the gradient of the pressure from the outer layer.

Another estimate is possible, assuming a logarithmic wind profile in the inner layer:

$$u(z) = \frac{\tau^{1/2}}{\kappa} \ln\left(\frac{z}{z_0}\right) \approx \left[\frac{u_*}{\kappa} + \frac{\Delta \tau}{2u_*\kappa}\right] \ln\left(\frac{z}{z_0}\right) , \tag{4.45}$$

with the value at the top determined from the outer layer solution: $u(l_t) = U_U + \Delta U(x, z = 0)$. Thus the shear stress perturbation results:

$$\Delta \tau = \frac{2\kappa u_*}{\ln(l_t/z_0)} \Delta U , \tag{4.46}$$

where (4.22) or (4.23) can be used to eliminate the logarithm. This expression suggests that the shear stress perturbation in the upper part of the inner layer is in phase with the pressure perturbation, thanks to (4.38). The ambiguity cannot be solved simply: note that earlier studies (Hunt and Richards 1984) suggested that the shear stress is approximately in phase with the horizontal pressure gradient if the inner layer is modelled as a laminar layer, while it is in phase with the pressure in the turbulent case. Anyway, negative minima of the shear stress perturbation indicate (qualitatively) that the total shear stress reduces: small shear stress means

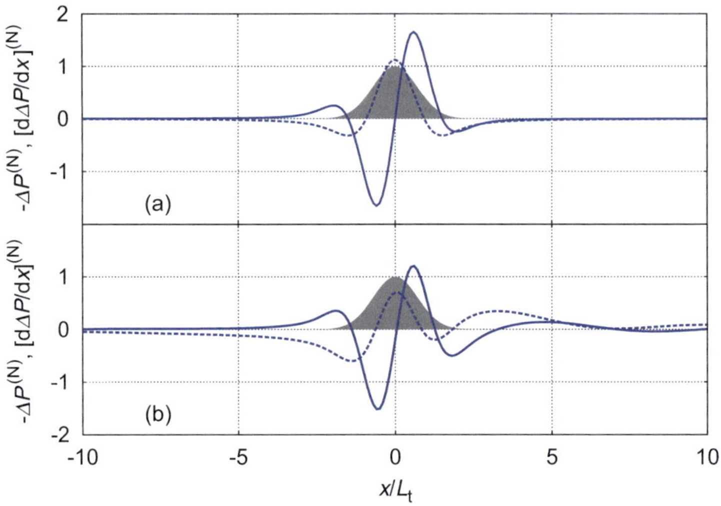

Fig. 4.6 Normalized perturbation to the pressure (*dashed line*) and to the pressure gradient (*continuous line*) alongwind at $z = 0$ from the outer layer solution. (**a**): $F_{\mathrm{I}}^{-1} = 0$; (**b**): $F_{\mathrm{I}}^{-1} = 1$

small mean velocity gradient, and thus a trend (extrapolating the linear solution) towards conditions of no wind at all. A region where the mean velocity and the shear stress approach zero is a region where the vertical exchanges are inhibited (on the average): it is a separated region (Tampieri 1987).

In Fig. 4.6 the negative of the pressure and the pressure gradient along x are reported for two stability conditions. The normalization is obtained by dividing the pressure by $\rho_{00} U_{\mathrm{U}}^2 H_{\mathrm{t}}/L_{\mathrm{t}}$. The analysis suggests that the shear stress perturbation has local (negative) minima upstream and in the lee of a hill: this broadly indicates the locations where separation can occur, as discussed in the following Section. This example shows also that the minimum upstream increases in amplitude with increasing stability: upstream separation is expected to be more evident in stable flows.

4.3.5 *The Turbulent Wake and the Separation*

Downwind of an obstacle, turbulence is modified. The large eddies shed by the obstacle are advected, and possibly decay with the distance: a turbulent wake occurs. The horizontal size of the eddies is influenced by the obstacle dimensions. Close to the foot of the downwind slope (in neutral conditions), for large enough slopes, separation can be observed. Note that separation refers to a unsteady process.

Figure 4.7 displays a flow visualization from a laboratory valley: the single realization (upper panel) shows an irregular motion, and the fluid parcels are thus expected to go in to and out from the separated region. Taking a number of images

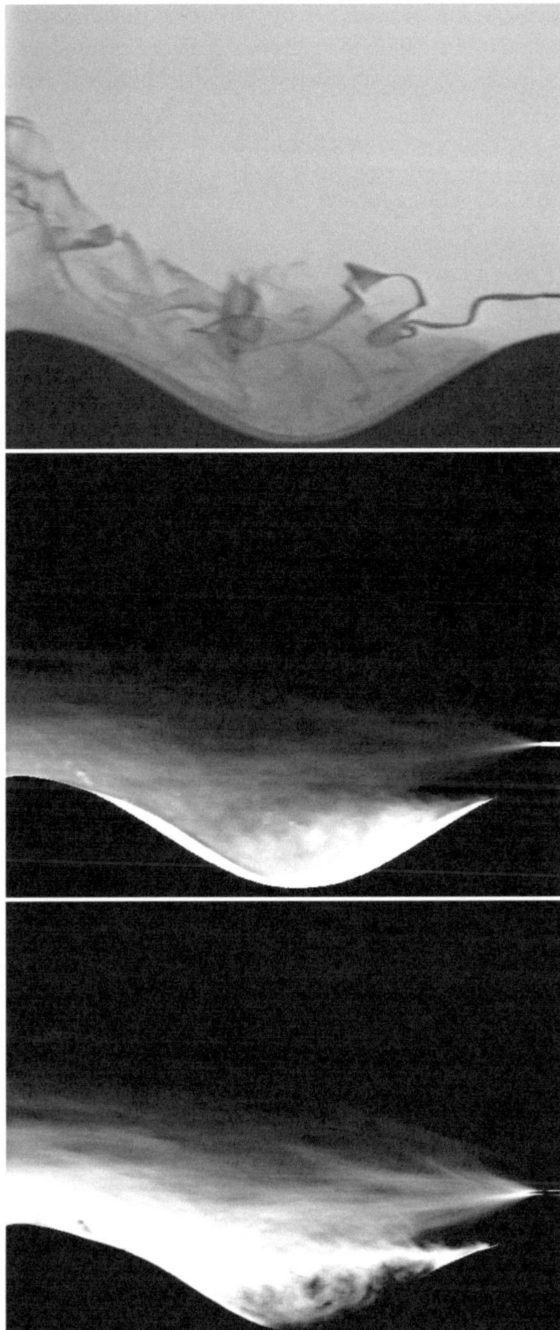

Fig. 4.7 Laboratory visualization of the flow in the separated region. The flow is *from right to left*, the valley is 2D, and the visualization is obtained from the dye continuously emitted by the source (*visible at the right side*), and depicted in the *upper panel*, which is a still from a continuous recording. By filtering these images with a threshold value for the *gray level*, every pixel is characterized by the presence or absence of dye: the concentration is 1 or 0 in that pixel. The *intermediate image* is the average concentration based on 3000 images: it is the ensemble average. *Dark areas* correspond to negligible concentration, *white areas* to high concentration: the maximum values correspond to the separated region. The *bottom image* shows the variance of the concentration: the highest values are found approximately in correspondence with the boundary between free-flow and separated regions. Courtesy Alberto Maurizi, CNR ISAC, Italy

an ensemble average can be obtained: the middle panel shows that the average concentration attains a maximum in a volume near the upstream slope (in the separated region); the variance (lower panel) is largest at the boundary between the separated region and the free-flow region.

In the lee of a gentle hill the location of the separation is dominated by the minimum of the shear stress in the inner layer, while for a bluff body the separation is usually activated by the geometrical shape (see Sect. 4.6.4.1).

4.3.6 Spectra Modifications

The analysis about the time scales in the inner and outer layers can be extended to the velocity spectrum: high frequency (large wavenumber) components are expected to be in equilibrium with the local value of the momentum flux, while the low frequency ones are distorted by the modifications of the mean wind profile. Moreover, the curvature of the streamlines affects the large-scale part of the spectrum. A discussion of the main factors influencing the spectra over topography can be found in Frank (1996). These modifications are of special interest for engineering applications, because the impact on the structures is different for different eddy size (and periods): a simple model has been proposed by Mann (2000).

The impact of small topographic features is documented in the literature: see for instance Panofsky et al. (1982), Tampieri et al. (2003), Mammarella et al. (2005), Berg et al. (2011, Figs. 11 and 12) among others.

4.4 Subgrid Effects of the Heterogeneous Surface Features

For small scale heterogeneities (heterogeneities not treated explicitly), we apply the general description of the flow based on the similarity approach, and investigate its limits. A couple of examples are discussed here.

4.4.1 Distributions of Roughness Elements on a Flat Surface

Let consider a surface with a distributions of roughness elements of different characteristics. We want to determine the blending height h_b, that is the height at which the flow forgets its local features, and the characteristics of the area-averaged flow above it. In other words, the effects of the heterogeneities are recognizable up to the blending height, while above the mean wind profile is characterized by an effective roughness length z_{0e}, and by friction velocity and stability parameter consistent with the area-averaged properties of the surface. Note that the area-averaged profile is what a numerical model gives as output, making the difference with local measured values.

The effective roughness is expected to be the average of the different values (Taylor 1987; Mason 1988), possibly weighted on the basis of the fraction of area covered with the same elements:

$$\ln(z_{0e}) = \frac{1}{N}\sum_{i=1}^{N} \ln(z_{0i}) . \tag{4.47}$$

The blending height can be estimated as the height at which the vertical divergence of the momentum flux balances the advection, or the height at which the diffusive effects balance the advection: the arguments are the same that lead, respectively, to the relationships (4.23) and (4.15) for the inner layer depth for flow over topography, where the topography scale length is substituted by the roughness change scale length L_r, using (4.23):

$$\frac{h_b}{L_r}\left[\ln\frac{h_b}{z_{0e}}\right]^2 = 2\kappa^2 . \tag{4.48}$$

The numerical coefficient $2\kappa^2$ may be modified on the basis of observations and numerical simulations; for periodic streamwise patches L_r can be taken proportional to the period, while for a generic distribution of roughness elements can be derived from the correlation function of their space distribution.

Numerical simulations of flow over changing roughness highlight some general features (see, for instance Miller and Stoll 2013). A relevant aspect is the enhancement of the vertical exchange, with respect to the homogeneous surface: the blending height is smaller if L_r is smaller; moreover, the growth rate of the IBLs increases as L_r decreases. Also, the vertical profiles of the mean velocity are in equilibrium with the local momentum flux near the ground (the adjustment of the friction velocity is rapid: see Fig. 4.8, consistent with the observations made in Sect. 4.2.1), but the IBL which forms downstream of each roughness change is lower, at the same downstream distance, than the IBL of the case discussed in the Sect. 4.2.1 where $L_r \to \infty$.

4.4.2 *The Effective Roughness of Topography*

If the topographic features are small with respect to the grid size of a model, their interaction with the PBL must be parameterized. In such situation the topography exerts a drag on the atmospheric flow, and this drag is of particular interest for meteorological modelling. The most commonly used approach is the bulk similarity approach, consisting of using an effective roughness length and a similarity wind profile (see Brown and Wood 2003). Earlier studies (Wood and Mason 1993) used

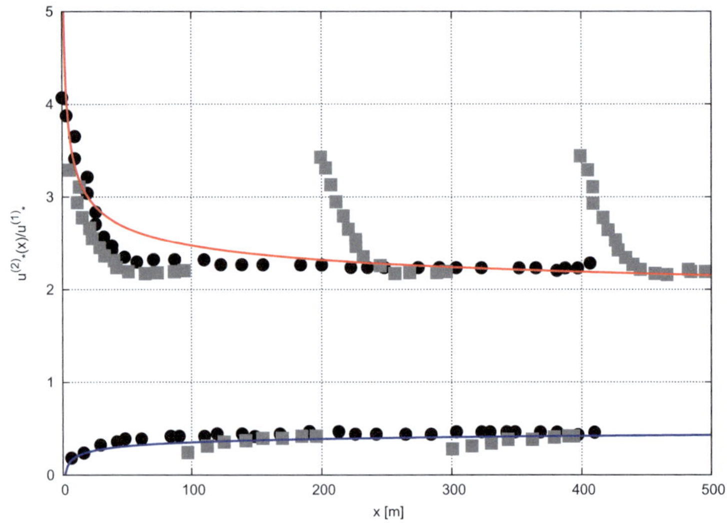

Fig. 4.8 Local variations of the friction velocity for periodic changes of roughness length, from the numerical simulations by Miller and Stoll (2013). Two stripes of different roughness (z_{01} = 0.1 m, z_{01} = 0.001 m), i.e. a 2D configuration, are placed normal to the mean wind; periodic conditions are simulated. Two experiments are documented: for the first (*black circles*) the width is 100 m and for the second (*gray squares*) it is 400 m. The ratio of the friction velocities upstream and downstream of each change in roughness is well represented by (4.5) using for h_i (4.7): *red line*: change from z_{02} to z_{01}, *blue* from z_{01} to z_{02}. Locally the adjustment to the local roughness is fast (faster than the model prediction)

simplified topographies, most recently multi-scale topography has been investigated using LES (Wan and Portè-Angel 2011).

A simple estimate of the contribution of the sub-grid scale topography (i.e. the length scale of the topography is smaller than the grid size of the model) to the effective roughness is to assume that it is proportional to the sub-grid standard deviation of the topographic elevation field $H_t f(x, y)$:

$$z_{0e}^2 = CH_t^2 \langle \overline{f^2} - \overline{f}^2 \rangle + z_0^2 \,, \tag{4.49}$$

below a critical value of the topography variance, while for larger values of the topography variance the quantity $z_{0e}^2 - z_0^2$ tends to saturate. Here z_0 is the roughness of the surface (constant in the area of averaging, or in general the effective roughness of the flat surface, see Sect. 4.4.1), and C is a function of z_0, estimated to be $10^{-4} \pm 10^{-3}$ for realistic values of the parameters (Wan and Portè-Angel 2011). Thus, the effective roughness is related to the variance of the topography and to the surface roughness.

4.5 Low Wind, Small Vertical Fluxes

As the wind is below a threshold, say $\sim\ 1\,\mathrm{m\,s^{-1}}$, the dynamics of the PBL departs from the paradigmatic conditions and the general rules described in Chap. 3. Although the expression "low-wind condition" is common, it is important to note that the departure from the ideal conditions occurs because the forcing reduces (Sect. 4.1.1). As far as "low-wind" corresponds to conditions of low shear stress production of TKE, and more generally to reduced dynamic effects, the stability do not affects, broadly speaking, the observed phenomenology. For instance, we observe deviations from similarity under very stable conditions (low wind, small vertical fluxes), while the global structure of the CBL driven by the heat flux (as described in Sect. 3.6) is to a large extent independent on the intensity of the wind. Note that separation is associated with low winds. The low-wind conditions are relevant in the parameterizations for dynamics (weather) and dispersion modelling, so deserve special attention also in applicative contexts.

Some basic observations are summaried here:

- the wind direction variability increases as the wind intensity decreases (Joffre and Laurila 1988) leading to difficulty in the identification of a mean direction for the wind; the occurrence of meandering, i.e. relatively low frequency oscillations of the wind direction (Mortarini and Anfossi 2015), emphasizes the critical dependence of the results from the time interval used for averaging the observations;
- the observations of the momentum flux, of the lateral and vertical variances of the velocity show a broad dependence on the mean wind velocity. For instance, the linear relation linking the momentum flux and the mean velocity derived by Maryon (1998) for a set of data including a full range of wind velocities represents the overall behaviour but misses the complexities appearing at low wind. The argument is under discussion in the literature. Some observations from CCT and CABAUW are reported in Fig. 4.9 together with the curves from Mahrt et al. (2015, Fig. 3b): it is evident that the variability at about $U\ <\ 1\,\mathrm{m\,s^{-1}}$ is extremely large, but the examined quantity for every data set levels at a value which is broadly constant. Correspondingly, the variances of the velocity components normalized with u_*^2 do not present any very clear dependence from the wind speed, see for instance Moraes et al. (2005, p. 3107), or Trini Castelli and Falabino (2013);
- a similar analysis can be carried out for the temperature variance and the temperature scale. The temperature variance reduces without going to zero in quasi-neutral conditions: as noted before (Sects. 3.4.3.4 and 4.1.1) small scale heterogeneities can produce non-negligible temperature fluctuations. It is also important to notice that the temperature scale is proportional to the temperature difference, according to the QNBL relation:

$$-\,\theta_* \approx \frac{\kappa}{\ln(z_2/z_1)}\left[\Theta(z_2)-\Theta(z_1)\right]\,, \tag{4.50}$$

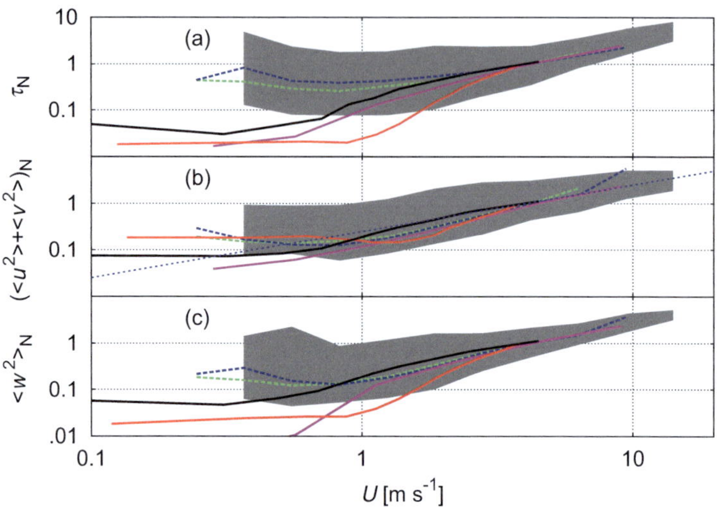

Fig. 4.9 (**a**) Momentum flux, (**b**) horizontal velocity variance and (**c**) vertical velocity variance, normalized to their respective values at $U = 4\,\mathrm{m\,s^{-1}}$, for different stabilities and data sets. *Gray shaded area*: QNBL from CCT (includes data within 10 to 90 percentiles); *green dashed*: CBL ($z/L < -1$) from CCT; *blue dashed*: SBL ($z/L > 1$) from CCT; *pink continuous line*: QNBL from CABAUW; *red continuous line*: stable data, redrawn from Mahrt et al. (2015); *black continuous line*: all stabilities, redrawn from Trini Castelli and Falabino (2013, Fig. 9, GRAZ data). *Blue dotted line*, in (**b**): from Maryon (1998, p. 119)

and thus goes to zero for zero temperature gradient while the temperature fluctuations variance is larger than zero in the entire temperature difference range, as can be seen in Fig. 4.10.

- in quasi-neutral conditions, the logarithmic wind profile is the result of a balance with the momentum flux: in low-wind conditions we expect departures from the logarithmic profile because the flux diminishes, and in fact the wind profile can be non-monotonic. However, we can use a logarithmic profile as a rough, but robust, approximation, and we can consistently derive the friction velocity $u_*^{(p)}$ from the mean wind measurements:

$$u_*^{(p)} = \frac{\kappa\,[U(z_2) - U(z_1)]}{\ln(z_2/z_1)} \ , \tag{4.51}$$

(the same can be made making a fit with more than two observations) and compare with the friction velocity measured by a sonic anemometer, $u_*^{(s)}$, possibly at an intermediate height. The ratio $u_*^{(s)}/u_*^{(p)}$ is plotted in Fig. 4.11: while it is not far from unity for wind velocity larger than, say, $2\,\mathrm{m\,s^{-1}}$, for lower velocities the spread of the data increases, and the profile departs from the logarithmic one, especially for the lower instrument (red circles in the Figure): the actual difference between the velocities measured at the two levels is smaller

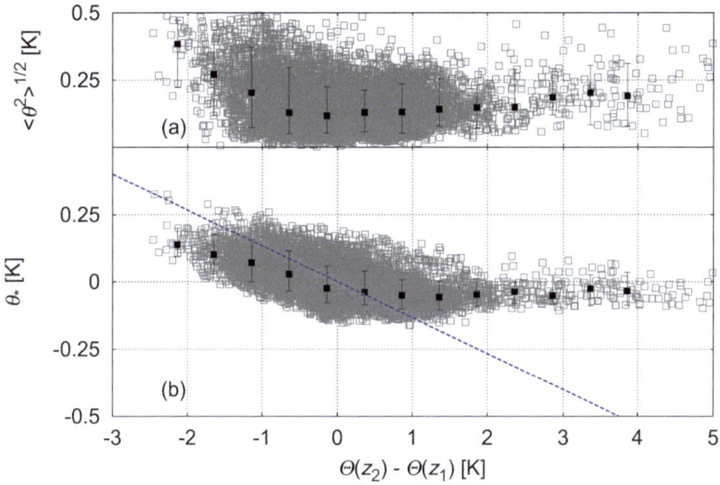

Fig. 4.10 (**a**) Standard deviation of the temperature variance; (**b**) temperature scale as function of the mean temperature difference, from CABAUW data. Here $z_1 = 3\,\text{m}$, $z_2 = 60\,\text{m}$, $\langle\theta^2\rangle$ and θ_* are measured at 3 m. The *dashed line* represents the relation between the temperature difference and the temperature scale described by the logarithmic law (the quasi-neutral limit)

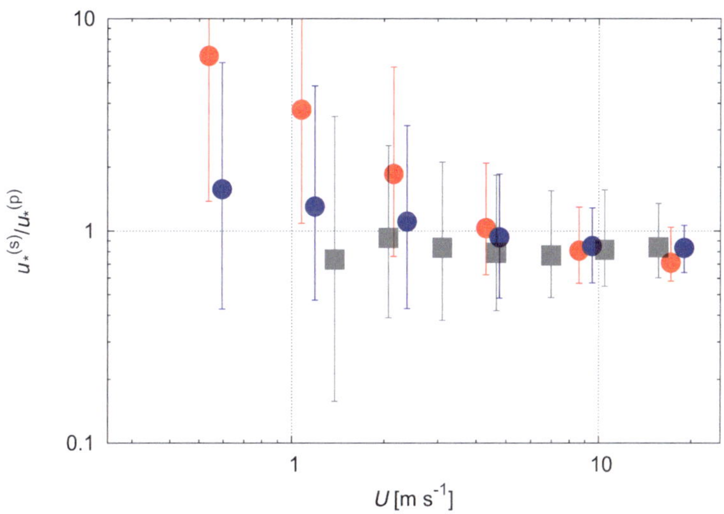

Fig. 4.11 Ratio between the measured (at level z) and the estimated from the mean wind profile, from (4.51), friction velocity as function of the mean wind in the layer z_1, z_2, for QNBL: here $|z/L| < 0.05$. *Red dots*: CCT data, with $z_1 = 2\,\text{m}$, $z_2 = 4.8\,\text{m}$, $z = 3.7\,\text{m}$; *blue dots*: CCT data, with $z_1 = 4.7\,\text{m}$, $z_2 = 10.3\,\text{m}$, $z = 7.5\,\text{m}$; *gray squares*: CABAUW data, with $z_1 = 10\,\text{m}$, $z_2 = 20\,\text{m}$, $z = 3\,\text{m}$

than that expected from the measured value of the friction velocity. This result can be rephrased in terms of flux-gradient relationship, and means that the actual eddy diffusion coefficient is larger than that expected from similarity $\kappa u_* z$; this result is qualitatively analogous to the larger eddy diffusion coefficient that result in the perturbed inner layer over an hill, see (4.20). Note also that the value of $u_*^{(s)}/u_*^{(p)}$ depends on the choice of the value of the von Karman constant: the data in the Figure would suggest an estimate $\kappa \approx 0.35$ instead of the value adopted here of 0.4, for high winds. These observations suggest that departures of κ from this range could arise if data with different velocities, including low ones, are mixed (see for a discussion about the von Karman constant Wieringa 1980; Andreas et al. 2006).

- the extension of the similarity relations (3.58) and (3.61) to the full range of wind velocity is of course quite appealing, at least for modeling purposes. As far as the velocity components are concerned, there are contrasting results in the literature, affected by the bias due to the normalization to the friction velocity, which becomes very small in low wind conditions. Furthermore, the averaging time affects the variances and the momentum flux in a different way: increasing the averaging time increases the variances (especially those of the horizontal components) because the meandering is included. At the same time, the meandering does not contribute to the vertical momentum flux, so that the normalized variances increase with the averaging time, as shown by Maryon (1998). The averaging time used by Anfossi et al. (2005) or Trini Castelli and Falabino (2013) is 1 h by instance, while the one used by Martins et al. (2009) or in the data presented in Fig. 3.9 is less or equal to 10 min: correspondingly, the normalized horizontal variances are larger for the former authors than for the last ones, in low wind conditions, while are consistent among all the authors, in high wind conditions.

4.6 Canopy Flow and the Urban PBL

Both the urban environment and a vegetative canopy are characterized by a distributed set of obstacles that interact with the wind and the temperature fields. In both cases we are interested in the flow above the canopy, which is seen as a set of roughness elements, and in the flow into it. Because the canopy is an heterogeneous environment, transitions and internal boundary layers are expected to occur in the flow above, and can be faced with the background of knowledge developed in the previous parts of the book. As far the flow into is concerned, we can be interested in its average behaviour (using averages over different areas, whose extension is relevant for the problem we have to tackle), and in specific and detailed aspects. The average behaviour can be tackled using some basic and general concepts; detailed description is mainly obtained by numerical simulations.

There is a huge literature on this topic: reference can be made to Roth (2000), Britter and Hanna (2003), Fernando (2010) and Fernando et al. (2010) for general reviews. Field experiments like URBAN (Allwine et al. 2002) or BUBBLE (Rotach et al. 2005) allowed an overall understanding of the PBL dynamics. Laboratory experiments (Castro et al. 2006) and field experiments with urban buildings simulated by regular arrays of identical obstacles (Zajiac et al. 2015; Roth et al. 2015) are useful to improve the understanding and the modelling of the interaction between wind and obstacles. A special attention has been paid to the thermal effects of the cities on the environment (heat island): see Grimmond and Oke (1999, 2002) or Rotach et al. (2005). Thermal effects of the buildings on the local PBL dynamics are relevant not only for the air circulation in streets, but also for the optimization of the energetic efficiency of the buildings and of the city as a whole.

The PBL over and in a city is thus characterized by many scales of forcing, and can be splitted into different layers. Broadly speaking:

- the layer above the building top: the urban boundary layer (UBL) develops as any other PBL over a rough surface;
- the layer in the building canopy, which can be considered as a porous medium: the urban canopy layer (UCL) is characterized by the average features of the flow, where the average is made on volumes which include both free air and obstacles;
- a matching layer between UBL and UCL, called the roughness layer (RS). At the top of this layer the effects of the surface heterogeneity disappear: its thickness is conceptually similar to the blending height discussed when dealing with the roughness heterogeneity;
- the flow around the buildings and in the streets.

Note that some of the problems we are facing are formally equivalent to those that arise in and above a tall vegetative canopy (a forest).

4.6.1 Some Scales and the Drag Due to the Buildings

To treat both UBL and UCL, we are interested in dealing with area-averaged quantities, not representative of a point, but of a (possibly homogeneous) area of the city.

Let consider a schematic city: Fig. 4.12.

Consider an area A over which we want to define the average features of the flow. This area is characterized by

- the mean building height

$$h_c = \frac{\sum_n h_{cn} A_{pn}}{\sum A_{pn}} \; ; \tag{4.52}$$

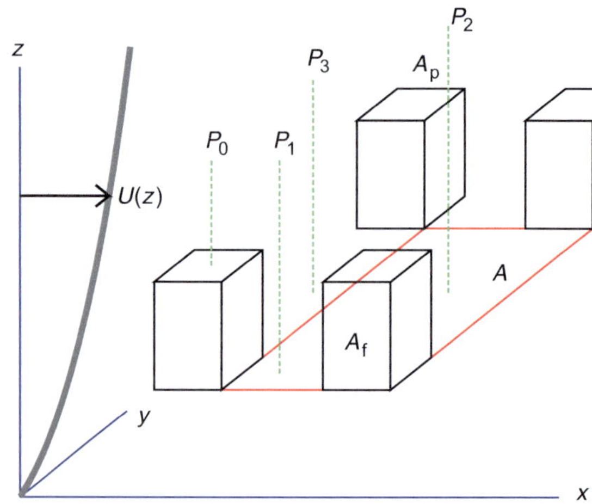

Fig. 4.12 Scheme for the density definitions: A_f is the frontal area of each building, A_p is the planar area, A is the area of averaging to obtain the average properties of the canopy. The positions indicated with $P_1, \ldots, P_4$ refer to the measurements by Castro et al. (2006), in part presented in Fig. 4.13

- the planar density

$$\lambda_p = \frac{\sum_n A_{pn}}{A} , \tag{4.53}$$

so that the volume of the urban canopy in the area A is $h_c A \lambda_p$;

- the frontal density

$$\lambda_f = \frac{\sum_n A_{fn}}{A} . \tag{4.54}$$

For a set of cubes, $\lambda_p = \lambda_f$.

Spatial changes of the characterizing parameters are obviously typical of the real cities. Thus a horizontal scale of variation in the direction x_i can be defined (Belcher et al. 2003):

$$L_c \approx \frac{\overline{U}^2}{|U_j(\partial U_i/\partial x_j)|} , \tag{4.55}$$

where $\overline{U}$ is a reference advection velocity in the layer. This definition grounds on the hypothesis that the reduction of velocity for a flow entering in a canopy, measured by the advection term, is due to the pressure (that scales with the velocity squared) changing on a distance L_c. This scale can be used to quantify any change of the canopy characteristics.

4.6.1.1 Averaging the Velocity into the Canopy

The mean wind averaged over a horizontal surface including obstacles is a measure of the mean flow into the canopy. The mean velocity will be decomposed in an averaged quantity plus a fluctuation about it: $U_i = \langle U_i \rangle + U_i'$ where now the brackets indicate the averaging over the area A. Of course, $\langle U_i' \rangle = 0$. The same holds for any scalar, in particular for the pressure.

In order to obtain the equations for the area averaged mean wind, the Reynolds equations must be (area) averaged consistently. As noticed by Wilson and Shaw (1977), Raupach and Shaw (1982) and Finnigan (2000), area average and derivation with respect to a spatial coordinate not always commute, because the domain inside A is multiply connected (i.e., it contains obstacles). This introduces new terms, which are synthetized as drag terms, which derive from the (viscous) interaction of the flow with the obstacles, and must be accounted for in the formulation of the momentum equation: see the Exercise 4.1.

Thus, neglecting the buoyancy, rotation and viscosity terms, the horizontal momentum equation for the area-averaged velocity reads:

$$\frac{\mathrm{D}\langle U_i \rangle}{\mathrm{D}t} = -\frac{1}{\rho_{00}} \frac{\partial \langle P \rangle}{\partial x_i} - D_i + \frac{\partial \tau_{ij}}{\partial x_j} , \tag{4.56}$$

where τ_{ij} is the negative of the area-averaged tensor of the turbulent fluxes, the material derivative is written in terms of the area-averaged velocity and D_i is the sum of the form and viscous drags due to the presence of the obstacles:

$$D_i = \frac{1}{\rho_{00}} \left\langle \frac{\partial P}{\partial x_i} \right\rangle - \nu \left\langle \frac{\partial^2 U_i}{\partial x_j \partial x_j} \right\rangle , \tag{4.57}$$

which shall be parametrized (see, for instance, Coceal and Belcher 2005).

4.6.1.2 The Canopy Drag

The definition of the horizontal scale of variation (4.55) allows to parameterize the drag (force per unit mass) exerted by the canopy elements on the flow as follows:

$$D_i = \frac{\langle U_i \rangle |\langle U_i \rangle|}{L_c} . \tag{4.58}$$

To find an expression for the term ρD_i (force per unit volume), let define a drag coefficient c_d in order to express the force per unit volume in the layer $(z, z + \Delta z)$ due to the presence of the buildings as $\rho_{00} \langle U_i \rangle |\langle U_i \rangle| c_d \sum_n \Delta A_{fn}$ where ΔA_{fn} is the frontal area in the layer of building n.

If the buildings are parallelepiped of mean height h_c it results $\Delta A_{fn} = A_{fn} \Delta z / h_c$. The fluid volume in the layer is $(1 - \lambda_p) A \Delta z$.

Thus:

$$\rho_{00} D_i = \frac{\rho_{00} \langle U_i \rangle |\langle U_i \rangle| c_d \sum A_{fn}}{(1-\lambda_p) A h_c} = \frac{\rho_{00} \langle U_i \rangle |\langle U_i \rangle| c_d \lambda_f}{(1-\lambda_p) h_c} \tag{4.59}$$

which leads us to express the horizontal scale as function of the canopy scales

$$L_c = \frac{1}{c_d} \frac{1-\lambda_p}{\lambda_f} h_c \,. \tag{4.60}$$

Approximately, $c_d \simeq 1$ (Coceal and Belcher 2004) independent on height.

4.6.2 The Flow Above the Canopy

In the Urban Boundary Layer (UBL) the surface layer similarity holds, providing that the surface can be approximated by an homogeneous surface with proper equivalent quantities. A discussion on the extensions of similarity, especially oriented to weather modelling, is presented by Baklanov et al. (2005).

In neutral conditions, the logarithmic profile (3.33) is observed (see Roth 2000, Fig. 1); the presence of internal boundary layers can be recognized for instance in the measurements presented by Pelliccioni et al. (2012, Figs. 22 and 23). Other observations confirming the consistency with the similarity theory are reported by Wood et al. (2010) or by Quan and Hu (2009) as far second-order moments are concerned. Note that the profiles have to be understood as an average over an area: they should be written in brackets.

The mean flow in the UBL is summarized in the Fig. 4.13, while second-order moments are reported in the Fig. 4.14, for quasi-neutral conditions. It is interesting to observe the overall superposition of the atmospheric and wind tunnel data, stressing the usefulness of both the approaches, and also the differences among the various data sets. Note also the increase of the momentum flux at $z \approx h_c$: see Fig. 4.14a.

The effects of stratification do not show departures from (do not add essential informations to) the similarity results discussed in Chap. 3, and are not reported here (see, for instance Roth 2000, Figs. 4, 5, 6, and 9).

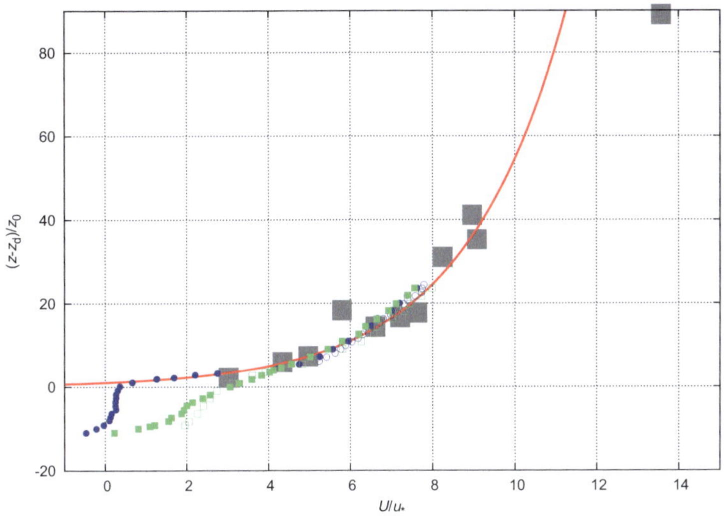

Fig. 4.13 Non-dimensional profiles of the mean velocity as function of the non-dimensional height. *Gray symbols*: redrawn from Roth (2000), referring to different cities; *blue and green symbols*: redrawn from Castro et al. (2006), referring to a wind-tunnel experiment and different positions (*open blue circles*: P_0; *full blue circles*: P_1; *full green squares*: P_2, *open green squares*: P_3) as indicated in Fig. 4.12. The wind tunnel data differ in the UCL, but collapse on the logarithmic law outside, consistent with the field observations. The *red line* gives the UBL (logarithmic) profile (3.33)

4.6.3 *The Average Flow in a Volume with an Array of Solid Obstacles: The Urban Canopy Layer*

Here we concentrate on the average flow within the canopy, neglecting local features but accounting for the overall behaviour. This approach makes sense for any problem where the integrated effect of the canopy (a city, a vegetative canopy) is investigated: for instance for modelling purposes.

4.6.3.1 Some Observations of the Flow in an Array of Obstacles

Observations in wind tunnel (for instance, Castro et al. 2006) or in specially arranged outdoor facilities (for instance, Roth et al. 2015), and proper analysis of measurements in real city environments, allow to evaluate the main features of the flow in UCL:

- the local profiles of the mean wind are affected by the position with respect to the obstacles: notable differences are found in urban canyons aligned with the wind with respect to cross-wind ones;
- wind reversal can be found downstream of buildings (separation);

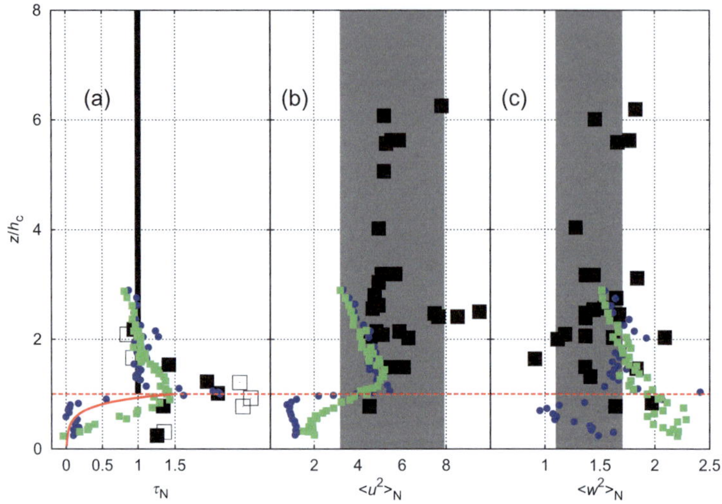

Fig. 4.14 Modulus of the momentum flux (**a**), of the streamwise velocity component variance (**b**) and of the vertical one (**c**) from the measurements by Castro et al. (2006): *blue* (P_1) and *green* (P_2) *symbols*, and from the observations reviewed by Rotach et al. (2005), *gray and black symbols* in (**a**), and by Roth (2000), *black symbols* in (**b**) and (**c**), as function of z/h_c. All the second-order moments are normalized to the friction velocity squared, as measured in the (almost) constant flux layer (for $z/h_c > 1.5$). The *gray* areas in (**b**) and (**c**) highlight the variability of the ratios from the surface layer values of Table 3.3. The *red continuous line* corresponds to (4.65), properly normalized, with $a = 2.9$, not far from the value suggested by Castro et al. (2006). The *red dashed line* evidences the obstacle mean height

- the wake of the buildings interacts, as well as eddies shed by the obstacles, leading to a very complex environment;
- the momentum flux has a maximum around the obstacle height, and then decreases towards the ground, with different rates depending on the characteristics of the canopy;
- similarly, there is an intense production of TKE at heights $\sim h_c$.

Some experimental mean wind profiles are reported in Fig. 4.15 together with the model profile (4.63) in the UCL. From the observations, an estimate of the area averaged values of wind and momentum flux can be derived, and compared with models.

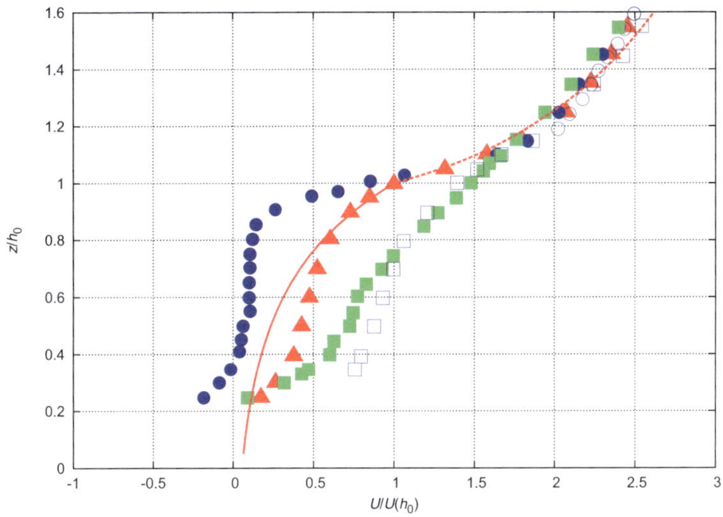

Fig. 4.15 Non-dimensional profiles of the mean velocity as function of the non-dimensional height in the UCL. Data are redrawn from Castro et al. (2006), referring to a wind-tunnel experiment and different positions (*open blue circles*: P_0; *full blue circles*: P_1; *full green squares*: P_2, *open green squares*: P_3) as indicated in Fig. 4.12. The *red triangles* represent the average (the experimental estimate of the area averaged profile, according to the definition given in the text). The *red continuous line* gives the UCL (exponential) profile (4.63), the *dashed line* the logarithmic profile (3.33)

4.6.3.2 A Model for Mean Wind and Momentum Flux

In steady horizontally homogeneous conditions, and considering the mean wind aligned with the x-axis, (4.56) reduces to :

$$\frac{\mathrm{d}\tau_{13}}{\mathrm{d}z} - D_1 = 0\,, \tag{4.61}$$

where the momentum flux can be parameterized with a mixing-length closure (2.84): $\tau_{13} = l_c^2(\mathrm{d}\langle U_1\rangle/\,\mathrm{d}z)^2$ and the drag is given by (4.58). The equation for the mean velocity reads:

$$\frac{\mathrm{d}}{\mathrm{d}z}\left[l_c^2\left(\frac{\mathrm{d}\langle U_1\rangle}{\mathrm{d}z}\right)^2\right] = \frac{\langle U_1\rangle^2}{L_c}\,. \tag{4.62}$$

The solution for l_c constant with height reads (Cionco 1965; Macdonald 2000):

$$\langle U_1 \rangle(z) = \langle U_1 \rangle(h_c) \exp\left[a\left(\frac{z}{h_c} - 1\right)\right] , \text{ for } z \leq h_c , \tag{4.63}$$

where

$$a^3 = \frac{h_c^3}{2l_c^2 L_c} , \tag{4.64}$$

is called attenuation factor (or density of the urban canopy). Note that if $l_c \propto h$ then $a^3 \propto h_c/L_c$.

The momentum flux profile reads, from the adopted closure:

$$\tau_{13} = l_c^2 \left(\frac{d\langle U_1 \rangle}{dz}\right)^2 = \langle U_1 \rangle(h_c)^2 l_c^2 \frac{a^2}{h_c^2} \exp\left[2a\left(z/h_c - 1\right)\right] . \tag{4.65}$$

More detailed models (Coceal and Belcher 2004; Kono et al. 2010) suggest that the mixing length changes with height, leading to more complex forms of the profiles.

4.6.4 Heterogeneous Urban Canopy

We are still considering the area averaged properties of the flow, but there is a practical interest in dealing with changes of these properties in the urban environment: more specifically, in changes of h_c and of L_c. The problem must be treated in general via numerical simulations with models of different complexity. A background assessment can be found in Belcher et al. (2003) and Coceal and Belcher (2004) and reviewed by Zajiac et al. (2015, Sect. 2.2 in particular), in terms of linearized perturbations.

A simple reasoning leads to understand the basic effects on the mean wind due to the changes of the canopy characteristics. Let consider two equilibrium conditions, over the upstream canopy (index 1) and the downstream canopy (index 2). At a level larger than the maximum building heights ($z_r \gg \max[h_{c1}, h_{c2}]$) the wind takes a reference value U_r constant with the downstream distance (the same concept of the wind above the blending height, see Sect. 4.4.1). Assume a the logarithmic profile (3.33) from z_r down to h_c, and the exponential one (4.63) from h_c to the ground. The resulting mean wind profile over the canopy i is characterized by h_{ci}, L_{ci}, $z_{di} = \alpha(a)h_{ci}$, $z_{0i} = \beta(a)h_{ci}$. To ensure continuity of the mean wind gradient at the matching height, some further conditions have to be applied: see Exercise 4.2. (In a even simpler view, the displacement height and the roughness length are assumed to be related only to the obstacle height.)

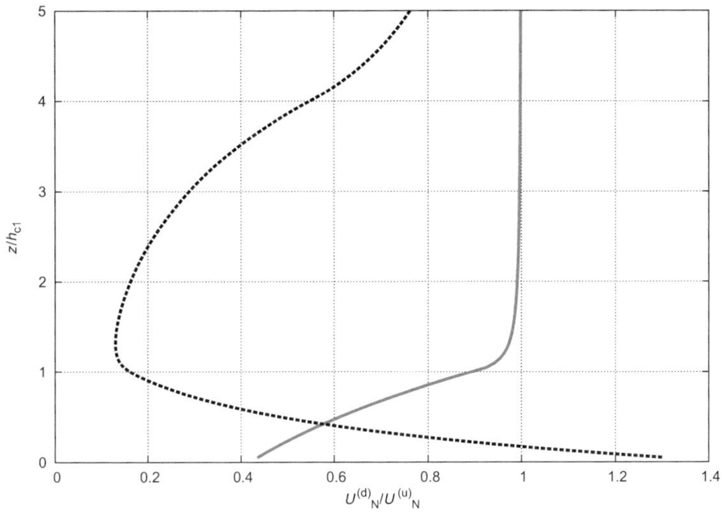

Fig. 4.16 Ratio of the downwind $U_N^{(d)}$ to the upwind $U_N^{(u)}$ mean wind profiles normalized to a reference value (here, the mean wind at $10h_{c1}$) for a change from a less dense to a denser canopy within the same building height (*continuous gray line*) and for a change from a lower to a taller canopy, with the same density (*dashed black line*), according to the model of Exercises 4.2 and 4.3

We consider the equilibrium wind profiles for the two canopies, normalized to the reference value U_r, and compute their ratios for changes of canopy features: see Exercise 4.3 for some exemplar cases.

Thus, considering a change from a less dense to a denser canopy with the same height: $h_{c1} = h_{c2}$, $L_{c1} > L_{c2}$ so that $a_1 < a_2$, the ratio of the downstream mean wind to the upstream one $U_2(z)/U_1(z) \leq 1$: obviously, the flow slows down in the transition (Coceal and Belcher 2005, case 1), as can be seen in Fig. 4.16. Considering the change from a lower to a taller canopy with the same density: $h_{c1} < h_{c2}, a_1 = a_2$ so that $L_{c1} < L_{c2}$, it results that $U_2(z)/U_1(z) \leq 1$ for $z \geq h_{c1}$, but near the surface the downstream wind increases with respect to the upstream one $U_2(z)/U_1(z) > 1$ for $z \ll h_{c1}$: see Fig. 4.16 (Coceal and Belcher 2005, case 2). In words, a downstream taller canopy with the same density of the upstream one leads to enhancement of the wind near the ground (and the reverse occurs in the change from a taller to a lower canopy: the flow near the surface slows down).

4.6.4.1 The Flow Around Buildings and the Urban Canyons

The flow in the space left among buildings is affected by the features of the buildings and by the presence of elements like trees in the roads. The heating/cooling effects are relevant and related to the radiation and the details of the surfaces (walls, roofs, roads). Numerical and experimental studies describe specific cases: for instance, measurements of flow and dispersion are reported by Soulhac et al. (2009), Finn et al. (2010), Li et al. (2012, 2015) and Klein and Galvez (2015), while numerical simulations are reported by Cheng and Liu (2011).

Exercises

4.1 Show that area averaging and space derivatives do not commute in general for multiply-connected domains.

4.2 Discuss the conditions to obtain matching between the mean wind into the UCL and the wind above it (in UBL) for quasi-neutral conditions.

4.3 Compute the nondimensional profiles of the mean wind in UCL and UBL (for quasi-neutral conditions), and the ratios of the normalized velocities, following and extending cases reported in Fig. 4.16.

References

K.J. Allwine, J.H. Shinn, G.E. Streit, K.L. Clawson, M. Brown, Overview of URBAN 2000. Bull. Am. Meteorol. Soc. **83**, 521–536 (2002)

E.L. Andreas, K.J. Claffey, R.E. Jordan, C.W. Fairall, P.S. Guest, P.O.G. Persson, A.A. Grachev, Evaluation of the von Karman constant in the atmospheric surface layer. J. Fluid Mech. **559**, 117–149 (2006)

D. Anfossi, D. Oettl, G. Degrazia, E. Ferrero, A. Goulart, An analysis of sonic anemometer observations in low wind speed conditions. Bound.-Layer Meteorol. **114**, 179–203 (2005)

M. Antonelli, R. Rotunno, Large-eddy simulation of the onset of the sea breeze. J. Atmos. Sci. **64**, 4445–4457 (2007)

A. Baklanov, P.G. Mestayer, A. Clappier, S.S. Zilitinkevich, S.M. Joffre, A. Mahura, N.W. Nielsen, On the parameterisation of the urban atmospheric sublayer in meteorological models. Atmos. Chem. Phys. Discuss. **5**, 12119–12176 (2005)

S.E. Belcher, N. Jerram, J.C.R. Hunt, Adjustment of a turbulent boundary layer to a canopy of roughness elements. J. Fluid Mech. **488**, 369–398 (2003)

J. Berg, J. Mann, A. Beckmann, M. Courtney, H.E. Jorgensen, The Bolund experiment, part I: flow over a steep, three-dimensional hill. Bound.-Layer Meteorol. **141**, 219–243 (2011)

E.F. Bradley, A micrometeorological study of velocity profiles and surface drag in the region modified by a change in surface roughness. Q. J. Roy. Meteorol. Soc. **94**, 361–379 (1968)

R.E. Britter, S.R. Hanna, Flow and dispersion in urban areas. Ann. Rev. Fluid Mech. **35**, 469–696 (2003)

A.R. Brown, N. Wood, Properties and parameterization of the stable boundary layer over moderate topography. J. Atmos. Sci. **60**, 2797–2808 (2003)

I. P. Castro, H. Cheng, R. Reynolds, Turbulence over urban-type roughness: deductions from wind-tunnel measurements. Bound.-Layer Meteorol. **118**, 109–131 (2006)

W.C. Cheng, C.-H. Liu, Large-eddy simulation of flow and pollutant transports in and above two-dimensional idealized street canyons. Bound.-Layer Meteorol. **139**, 411–437 (2011)

R.M. Cionco, Mathematical model for air flow in a vegetative canopy. J. Appl. Meteorol. **4**, 517–522 (1965)

O. Coceal, S.E. Belcher, A canopy model of mean winds through urban areas. Q. J. Roy. Meteorol. Soc. **130**, 1349–1372 (2004)

O. Coceal, S.E. Belcher, Mean winds through an inhomogeneous urban canopy. Bound.-Layer Meteorol. **115**, 47–68 (2005)

M. De Cervantes, *Don Quijote de la Mancha*, Florencio Sevilla Arroyo, modern edition, 2010 (1605)

H.J.S. Fernando, Fluid dynamics of urban atmospheres in complex terrain. Ann. Rev. Fluid Mech. **42**, 365–389 (2010)

H.J.S. Fernando, D. Zajiac, S. Di Sabatino, R. Dimitrova, B. Hedquist, A. Dallman, Flow, turbulence, and pollutant dispersion in urban atmospheres. Phys. Fluids **22**, 051301, 1–20 (2010)

D. Finn, K.L. Clawson, R.G. Carter, J.D. Rich, C. Biltoft, M. Leach, Analysis of urban atmosphere plume concentration fluctuations. Bound.-Layer Meteorol. **136**, 431–456 (2010)

J.J. Finnigan, Turbulence in plant canopies. Annu. Rev. Fluid Mech. **32**, 519–571 (2000)

H.P. Frank, A simple spectral model for the modification of turbulence in flow over gentle hills. Bound.-Layer Meteorol. **79**, 345–373 (1996)

J.R. Garratt, The internal boundary layer - a review. Bound.-Layer Meteorol. **50**, 171–203 (1990)

A.E. Gill, *Atmosphere-Ocean Dynamics* (Academic, San Diego, 1982)

C.S.B. Grimmond, T.R. Oke, Heat storage in urban areas: local-scale observations and evaluation of a simple model. J. Appl. Meteorol. **38**, 922–940 (1999)

C.S.B. Grimmond, T.R. Oke, Turbulent heat fluxes in urban areas: observations and a local-scale urban meteorological parameterization scheme (LUMPS). J. Appl. Meteorol. **41**, 792–810 (2002)

S.E. Gryning, E. Batchvarova, Analytical model for the growth of the coastal internal boundary layer during onshore flow. Q. J. Roy. Meteorol. Soc. **116**, 187–203 (1990)

J.C.R. Hunt, K. Richards, Stratified airflow over one or two hills. Bound.-Layer Meteorol. **30**, 223–295 (1984)

J. Hunt, K. Richards, P. Brighton, Stably stratified shear flow over low hills. Q. J. Roy. Meteorol. Soc. **114**, 859–886 (1988a)

J.C.R. Hunt, J.C. Kaimal, J.E. Gaynor, Eddy structure in the convective boundary layer - new measurements and new concepts. Q. J. Roy. Meteor. Soc. **114**, 827–858 (1988b)

P.S. Jackson, J.C.R. Hunt, Turbulent wind flow over low hills. Q. J. Roy. Meteorol. Soc. **101**, 929–956 (1975)

N.O. Jensen, E.L. Petersen, I. Troen, Extrapolation of mean wind statistics with special regard to wind energy applications, Technical Report WMO-WCP 86, World Climate Programme, WMO, Geneva, Switzerland, 1984

S.M. Joffre, T. Laurila, Standard deviations of wind speed and direction from observations over a smooth surface. J. Appl. Meteorol. **27**, 550–561 (1988)

L.H. Khurshudyan, W.H. Snyder, I.V. Nekrasov, Flow and dispersion of pollutants over two-dimensional hills: summary report on joint Soviet-American study. Technical Report EPA-600/4-81-067, EPA, Res. Tri. Pk., NC, 1981

L.H. Khurshudyan, W.H. Snyder, I.V. Nekrasov, R.E. Lawson Jr., R.S. Thompson, F.A. Schiermeier, Flow and dispersion of pollutants within two-dimensional valleys: summary report on joint Soviet-American study. Technical Report EPA Report 600/3-90/025, EPA 1990

P.M. Klein, J.M. Galvez, Flow and turbulence characteristics in a suburban street canyon. Environ. Fluid Mech. **15**, 419–438 (2015)

T. Kono, T. Tamura, Y. Ashie, Numerical investigations of mean winds within canopies of regularly arrayed cubical buildings under neutral stability conditions. Bound.-Layer Meteorol. **134**, 131–156 (2010)

X.-X. Li, R.E. Britter, L.K. Norford, T.-Y. Koh, D. Entekhabi, Flow and pollutant transport in urban street canyons of different aspect ratios with ground heating: large-eddy simulation. Bound.-Layer Meteorol. **142**, 289–304 (2012)

X.-X. Li, R.E. Britter, L.K. Norford, Transport processes in and above two-dimensional urban street canyons under different stratification conditions: results from numerical simulation. Environ. Fluid Mech. **15**, 399–417 (2015)

J. Lighthill, *Waves in Fluids* (Cambridge University Press, Cambridge, 1978)

D.K. Lilly, J.B. Klemp, The effects of terrain shape on nonlinear hydrostatic mountain waves. J. Fluid Mech. **94**, 241–261 (1979)

R.W. Macdonald, Modelling the mean velocity profile in the urban canopy layer. Bound.-Layer Meteorol. **97**, 25–45 (2000)

L. Mahrt, J. Sun, D.R. Stauffer, Dependence of turbulent velocities on wind speed and stratification. Bound.-Layer Meteorol. **155**, 55–71 (2015)

I. Mammarella, F. Tampieri, M. Tagliazucca, M. Nardino, Turbulence perturbations in the neutrally stratified surface layer due to the interaction of a katabatic flow with a steep ridge. Environ. Fluid Mech. **5**, 227–246 (2005)

J. Mann, The spectral velocity tensor in moderately complex terrain. J. Wind Eng. Ind. Aerodyn. **88**, 153–169 (2000)

C.A. Martins, O.L. Moraes, O.C. Acevedo, G.A. Degrazia, Turbulence intensity parameters over a very complex terrain. Bound.-Layer Meteorol. **133**, 35–45 (2009)

R.H. Maryon, Determining cross-wind variance for low frequency wind meander. Atmos. Environ. **32**, 115–121 (1998)

P.J. Mason, The formation of areally-averaged roughness lengths. Q. J. Roy. Meteorol. Soc. **114**, 399–420 (1988)

N.E. Miller, R. Stoll, Surface heterogeneity effects on regional-scale fluxes in the stable boundary layer: aerodynamic roughness length transitions. Bound.-Layer Meteorol. **149**, 277–301 (2013)

M. Miyake, Transformation of the atmospheric boundary layer over inhomogeneous surfaces, Science Report 5R-6, University of Washington, Seattle, USA, 1965

O.L. Moraes, O.C. Acevedo, G.A. Degrazia, D. Anfossi, R. da Silva, V. Anabor, Surface layer turbulence parameters over a complex terrain. Atmos. Environ. **39**, 3103–3112 (2005)

G.A.A. Moreira, A.A.C. dos Santos, C.A.M. do Nascimento, R.M. Valle, Numerical study of the neutral atmospheric boundary layer over complex terrain. Bound.-Layer Meteorol. **143**, 393–407 (2012)

L. Mortarini, D. Anfossi, Proposal of an empirical velocity spectrum formula in low-wind speed conditions. Q. J. Roy. Meteorol. Soc. **141**, 85–97 (2015)

S.C. Nanni, F. Tampieri, A linear investigation on separation in laminar and turbulent boundary layers over low hills and valleys. Il Nuovo Cimento C **8**, 579–601 (1985)

H. Panofsky, D. Larko, R. Lipschutz, G. Stone, E.F. Bradley, A.J. Bowen, J. Hojstrup, Spectra of velocity components over complex terrain. Q. J. Roy. Meteorol. Soc. **108**, 215–230 (1982)

A. Pelliccioni, P. Monti, C. Gariazzo, G. Leuzzi, Some characteristics of the urban boundary layer above Rome, Italy, and applicability of Monin–Obukhov similarity. Environ. Fluid Mech. **12**, 405–428 (2012)

L. Quan, F. Hu, Relationship between turbulent flux and variance in the urban canopy. Meteorol. Atmos. Phys. **104**, 29–36 (2009)

M.R. Raupach, R.H. Shaw, Averaging procedures for flow within vegetation canopies. Bound.-Layer Meteorol. **22**, 79–90 (1982)

D.P. Rogers, D.W. Johnson, C.A. Friehe, The stable internal boundary-layer over a coastal sea. 1. Airborne measurements of the mean and turbulence structure. J. Atmos. Sci. **52**, 667–683 (1995)

M.W. Rotach, R. Vogt, C. Bernhofer, E. Batchvarova, A. Christen, A. Clappier, B. Feddersen, S.E. Gryning, M.G., H. Mayer, V. Mitev, T.R. Oke, E. Parlow, H. Richner, M. Roth, Y.A. Roulet, D. Ruffieux, J.A. Salmond, M. Schatzmann, J.A. Voogt, Bubble - an urban boundary layer meteorology project. Theor. Appl. Climatol. **81**, 231–261 (2005)

M. Roth, Review of atmospheric turbulence over cities. Q. J. Roy. Meteorol. Soc. **126**, 941–990 (2000)

M. Roth, A. Inagaki, H. Sugawara, M. Kanda, Small-scale spatial variability of turbulence statistics, (co)spectra and turbulent kinetic energy measured over a regular array of cube roughness. Environ. Fluid Mech. **15**, 329–348 (2015)

S.A. Savelyev, P.A. Taylor, Internal boundary layers: I. height formulae for neutral and diabatic flows. Bound.-Layer Meteorol. **115**, 1–25 (2005)

W.H. Snyder, L.H. Khurshudyan, I.V. Nekrasov, R.E. Lawson Jr., R.S. Thompson, Flow and dispersion of pollutants within two-dimensional valleys. Atmos. Environ. A **25**, 1347–1375 (1991)

L. Soulhac, V. Garbero, P. Salizzoni, P. Mejean, R.J. Perkins, Flow and dispersion in street intersections. Atmos. Environ. **43**, 2981–2996 (2009)

F. Tampieri, Separation features of boundary-layer flow over valleys. Bound.-Layer Meteorol. **40**, 295–307 (1987)

F. Tampieri, I. Mammarella, A. Maurizi, Turbulence in complex terrain. Bound.-Layer Meteorol. **109**, 85–97 (2003)

P.A. Taylor, Comments and further analysis on effective roughness lengths for use in numerical three-dimensional models. Bound.-Layer Meteorol. **39**, 403–418 (1987)

P.A. Taylor, P.J. Mason, E.F. Bradley, Boundary-layer flow over low hills. Bound.-Layer Meteorol. **39**, 107–132 (1987)

S. Trini Castelli, S. Falabino, Analysis of the parameterization for the wind-velocity fluctuation standard deviations in the surface layer in low-wind conditions. Meteorol. Atmos. Phys. **119**, 91–107 (2013)

M.G. Villani, A. Maurizi, F. Tampieri, Discussion and applications of slab models of the convective boundary layer based on turbulent kinetic energy budget parameterisations. Bound.-Layer Meteorol. **114**, 539–556 (2005)

F. Wan, F. Portè-Angel, A large-eddy simulation study of turbulent flow over multiscale topography. Bound.-Layer Meteorol. **141**, 201–217 (2011)

J. Wieringa, A revaluation of the Kansas mast influence on measurements of stress and cup anemometer overspeeding. Bound.-Layer Meteorol. **18**, 411–430 (1980)

N.R. Wilson, R.H. Shaw, A higher order closure model for canopy flow. J. Appl. Meteorol. **16**, 1197–1205 (1977)

D.H. Wood, Internal boundary layer growth following a change in surface roughness. Bound.-Layer Meteorol. **22**, 241–244 (1982)

N. Wood, P.J. Mason, The pressure force induced by neutral, turbulent flow over hills. Q. J. Roy. Meteorol. Soc. **119**, 1233–1267 (1993)

C.R. Wood, A. Lacser, J.F. Barlow, A. Padhra, S.E. Belcher, E. Nemitz, C. Helfter, D. Famulari, C.S.B. Grimmond, Turbulent flow at 190 m height above London during 2006–2008: a climatology and the applicability of similarity theory. Bound.-Layer Meteorol. **137**, 77–96 (2010)

D. Zajiac, H.J.S. Fernando, M.J. Brown, E.R. Pardyjak, On flows in simulated urban canopies. Environ. Fluid Mech. **15**, 275–303 (2015)

O. Zeman, N.O. Jensen, Modification of turbulence characteristics in flow over hills. Q. J. Roy. Meteorol. Soc. **113**, 55–80 (1987)

Chapter 5
Turbulent Dispersion

Abstract This chapter deals with the problem of transport of tracers from a source. At first a few general results are discussed, highlighting the peculiar features of the turbulent dispersion, its similitudes to molecular diffusion and its differences. The statistical description allows to recognize the velocity-uncorrelated approach (the diffusion approach, valid for times larger than the characteristic time scale of the turbulent velocities) and the acceleration-uncorrelated approach (valid also for times shorter than the velocity time scale), recasting on formal bases the popular classification of Eulerian and Lagrangian dispersion models. The Lagrangian Stochastic Model (LSM) approach is discussed in detail, being the basis for many practical modeling tools, based on the use both of the observations and of the numerical simulations. Some aspects related to the problem of the fluctuations in the concentration of tracers are introduced. The inertial particle dispersion is treated shortly.

5.1 The Transport Problem for Fluid Parcels

The transport problem has been discussed in the Introduction. After the presentation of the atmospheric turbulence complexities, we are in a better position to understand that mixing of tracers needs to deal not only with the mean flow and the overall features of the microscopic (with respect to the spatial and temporal scales we are interested in) molecular diffusion process, but also with the continuous range of sizes of the eddies.

We face the problem by a statistical approach (the concentration being a synonymous of the probability density function of the parcel positions); in practice this approach is related to the space and time resolution of the observations or of the numerical simulations in use. Broadly speaking, if we know the wind averaged over times and lengths larger than the respectively integral scales, the entire spectrum of velocity will contribute to the statistically treated process. If finer resolution is attained, the motion is better resolved (the large eddies are explicitly described), and the statistical treatment is deserved to the unresolved part of the velocity field.

Hereinafter we will refer essentially to fluid parcels and the condition (2.2) holds. Note that this condition does not hold for inertial parcels (like aerosols), and some aspects are discussed in Sect. 5.7.

F. Tampieri, *Turbulence and Dispersion in the Planetary Boundary Layer*,
Physics of Earth and Space Environments, DOI 10.1007/978-3-319-43604-3_5

5.1.1 Probability Density Functions, Concentration and Well Mixed Condition (WMC)

The statistical bases for the formal treatment of the problem can be found in specialized books, like Gardiner (1990) or Risken (1989). A textbook addressing specifically environmental problem is Csanady (1973). Detailed discussion can be found also in Monin and Yaglom (1975). Numerical methods for stochastic equations are presented by Kloeden and Platen (1992).

We consider here the probability f that a fluid parcel moves from a state (characterized by a given position and velocity at a given time) to another one (again, position and velocity at a new time). The function f is a probability density function, with dimensions $\mathrm{m}^{-6}\mathrm{s}^{2}$. Note that if we neglect the velocity of the parcel, the state is characterized only by the position (in this case f has dimensions $\mathrm{m}^{-3}\mathrm{s}^{-1}$). In Sect. 5.6 we will deal with these two possibilities.

Note that the state may be characterized by the acceleration too, or by higher order derivatives of the position: a discussion is reported in Berloff and McWilliams (2002).

Let we consider parcels whose characteristics do not change in time (non reacting, non-decaying tracers) and without inertia. The concentration of the tracer parcels released from a source can be computed from the transition probability f. If $S(\boldsymbol{x}_0, \boldsymbol{v}_0; t_0)$ is the number of marked parcels in the initial position (in the state space) for time interval, volume and velocity range (namely, the source), then

$$n_S(\boldsymbol{x}, \boldsymbol{v}; t) = \int \mathrm{d}^3\boldsymbol{x}_0 \int \mathrm{d}^3\boldsymbol{v}_0 \int_{t_0<t} \mathrm{d}t_0\, f(\boldsymbol{x}, \boldsymbol{v}; t|\boldsymbol{x}_0, \boldsymbol{v}_0; t_0) S(\boldsymbol{x}_0, \boldsymbol{v}_0; t_0) \tag{5.1}$$

is the number of tracer parcels for volume and velocity range found in the final state at time t.

Then, the integration of n_S over the velocity gives the number of parcels for volume range, i.e. the average concentration of tracer released from the source S:

$$C_S(\boldsymbol{x}; t) = \int \mathrm{d}^3\boldsymbol{v} n_S(\boldsymbol{x}, \boldsymbol{v}; t) \tag{5.2}$$

If f is known, using (5.1) and (5.2) it is possible to compute the concentration of a tracer released by a source (the simplest example: an instantaneous point release with a given initial velocity). The knowledge of f is equivalent to the knowledge of the trajectories of each fluid parcel necessary to solve Eq. (2.1) and to compute the ensemble average of the positions.

For the present analysis, it is useful to define a special source, equal to the mean density of the fluid ρ_f (which is the probability of finding fluid parcels in a given volume, multiplied by the mass of the parcels themselves: in what follows, we shall avoid to introduce the mass just defining density as the number of parcels for unit volume) times the Eulerian probability density function of the velocity f_E:

$$S = \rho_f(\boldsymbol{x}_0, t_0) f_E(\boldsymbol{v}_0; \boldsymbol{x}_0, t_0)\delta(t_0) \tag{5.3}$$

According to the definition, this source is made by the fluid parcels leaving all the possible initial position, with initial velocity distributed as the flow velocity (namely, the Eulerian velocity distribution). If the integration in (5.1) is made over the entire space and velocity domain, then n_S must be the number (for volume and velocity range) of the fluid parcels in the final position, i.e. $n_S = \rho_f(\boldsymbol{x}, t) f_E(\boldsymbol{v}; \boldsymbol{x}, t)$. In words, if the source is given by parcels that sample the entire flow, their final distribution is the same as the initial one: parcel perfectly mixed at a given time will remain as such at all the times: the so-called Well Mixed Condition (WMC). In a slightly different form, this point was formulated by Novikov (1986) and Thomson (1987).

In a constant density fluid, substituting (5.3) in (5.1), namely, integrating over the initial time and over the entire space and velocity domain, we get

$$f_E(\boldsymbol{v}; \boldsymbol{x}, t) = \int d^3\boldsymbol{x}_0 \int d^3\boldsymbol{v}_0 f(\boldsymbol{x}, \boldsymbol{v}; t|\boldsymbol{x}_0, \boldsymbol{v}_0; t_0) f_E(\boldsymbol{v}_0; \boldsymbol{x}_0, t_0) \,. \tag{5.4}$$

This equation expresses the WMC in a constant density fluid, and will be applied to define the models for dispersion in the PBL.

A more general formulation which accounts for density variations can be found for instance in Thomson (1987), and is based on the conservation of the mixing ratio of the tracer in a compressible fluid: see Sect. 2.2.4, Eq. (2.16). Applied to a particular stochastic model, this point will be treated in Sect. 5.6.2.

5.2 Absolute Dispersion of Tracer Parcels

Absolute dispersion means that we apply the transport equation (2.1) to an ensemble of uncorrelated parcels: formally, each parcel belongs to a different realization of the flow. In steady conditions, parcels move independently each other if are released from the same point $\boldsymbol{x}_0$ at time intervals larger than the integral time scale; in homogeneous conditions, if released from a source larger than the integral length scale. In the following sections, the problem is treated in some idealized conditions.

5.2.1 *Steady Homogeneous Conditions: Taylor (1921)*

Let we study the statistics of fluid parcel position, described by (2.1) with the condition (2.2). Assuming a flow at rest: $U_i = 0$ and the source in the origin of coordinates: $x_i(0) = 0$, the mean position of the ensemble of parcels leaving the source at $t = 0$ is $\langle x_i \rangle_S(t) = 0$, while the second-order moments of the space coordinates read:

$$\langle x_i x_j \rangle_S(t) = \int_0^t \mathrm{d}t_1 \int_0^t \mathrm{d}t_2 \, \langle v_i(t_1) v_j(t_2) \rangle_S \; . \tag{5.5}$$

We shall use $\langle \; \rangle_S$ to represent the conditional average made on parcels leaving the source, not on the entire flow. When we substitute $\langle \; \rangle_S$ with $\langle \; \rangle$ (i.e., the conditional statistics with the Eulerian statistics), we imply that a large enough set of realizations is accounted for (this caveat will not explicitly stated hereinafter).

In homogeneous and steady conditions, we substitute $\langle v_i(t_1) v_j(t_2) \rangle_S$ with $\langle u_i(t_1) u_j(t_2) \rangle$, i.e. the covariance computed along the particle trajectory with the velocity covariance $R_{ij}(|t_2 - t_1|)$. The computation of the integral (5.5) is made in Exercise 2.6 and gives:

$$\langle x_i x_j \rangle_S(t) = 2 \int_0^t R_{ij}(s)(t-s) \, \mathrm{d}s \; . \tag{5.6}$$

For $t \ll T_{ij}$ the covariance is approximated by its (Eulerian) value at $s = 0$: $R_{ij}(0) = \langle u_i u_j \rangle$ and the ballistic regime occurs:

$$\langle x_i x_j \rangle_S(t) = \langle u_i u_j \rangle t^2 \; . \tag{5.7}$$

As the time increases, the growth rate of the position decreases by the effect of the reduction in the correlation. For $t \gg T_{ij}$

$$\begin{aligned} \int_0^t R_{ij}(s)(t-s) ds &= t \int_0^t R_{ij}(s) ds - \int_0^t s R_{ij}(s) \, \mathrm{d}s \\ &\simeq T_{ij} \langle u_i u_j \rangle t - C \; , \end{aligned} \tag{5.8}$$

so that the diffusive regime occurs:

$$\langle x_i x_j \rangle_S(t) \simeq 2 T_{ij} \langle u_i u_j \rangle t \; . \tag{5.9}$$

According to the definition (see, for instance Isichenko 1992, p. 1003), the diffusion coefficient of this process is:

$$D_{ij} \equiv \lim_{t \to \infty} \frac{\langle x_i x_j \rangle_S}{2t} = \langle u_i u_j \rangle T_{ij} \; . \tag{5.10}$$

Using the exponential model (2.112) for the covariance, the integral (5.6) can be done for any t and reads:

$$\langle x_i x_j \rangle_S(t) = 2 \langle u_i u_j \rangle T_{ij}^2 \left[\frac{t}{T_{ij}} - 1 + \exp\left(-\frac{t}{T_{ij}} \right) \right] \tag{5.11}$$

Observe that uncorrelated velocities are described by a δ-type covariance: for the u component $R_{uu}(s) = D\delta(s)$, where D is a constant of dimensions $m^2 s^{-1}$. Thus the integral (5.6) gives (Lighthill 1958, Eq. (10)):

$$\langle x^2 \rangle_S(t) = 2Dt \,, \tag{5.12}$$

i.e. the diffusive regime occurs for any t: δ-correlated velocities in turbulent flows occur for times larger than the Lagrangian time scale, obviously.

Using (2.56) written for time and frequency:

$$R_{ij}(s) = \int_0^\infty E_{ij}(\omega) \cos(\omega s) \, d\omega \tag{5.13}$$

in (5.6), the covariance of the coordinates can be written in terms of the energy spectrum:

$$\langle x_i x_j \rangle_S(t) = t^2 \int_0^\infty E_{ij}(\omega) \frac{\sin^2(\omega t/2)}{(\omega t/2)^2} \, d\omega = 2t \int_0^\infty E_{ij}(2k/t) \frac{\sin^2(k)}{k^2} \, dk \,, \tag{5.14}$$

where $k = 2\omega/t$.

Equation (5.14) contains a filter that selects which part of the spectrum is relevant for the covariance. It suggests that for small times (in the limit $t \to 0$) the entire spectrum is active, while for large times only the low frequencies give a significant contribution to the growth of the covariance.

Taylor (1921) formulation gives the paradigm for the turbulent dispersion at high Reynolds number (viscosity is neglected). The ballistic behaviour describes a phase in which the initial velocity of the parcel is relevant; the diffusion phase occurs when the velocities are uncorrelated in time, along each trajectory. The two phases are thus related to the intimate nature of the turbulence.

5.2.1.1 D_{ij} vs. K_τ

Both D_{ij} defined by (5.10) and K_τ defined by (2.80) and (2.84) in Sect. 2.4.6 are called eddy diffusion coefficients, and have the same form of a scale velocity times a scale length. The intimate difference is that K_τ (and the analogous for heat, or for any other scalar) is derived by imposing the existence of a flux-gradient relationship, and is thus an empirical closure of the equations necessary to eliminate the presence of high order moments; while D_{ij} derives from the intimate (statistical) nature of the turbulence and is thus theoretically based.

Note that D_{ij} indicates that turbulence can produce mixing by itself, while K_τ implies the presence of a mean gradient, necessary to have an eddy diffusion coefficient different from zero (in the discussion about CBL we had to face with non-local transport, i.e. mixing in the absence of gradient).

Furthermore, D_{ij} is related to a diffusive process in the true sense (the scale of the process is much larger than the scale of the motions that produce such process), while in the derivation of K_τ no hypothesis about the scales is made.

5.2.2 *Extension to Inhomogeneous Conditions*

The extension to inhomogeneous conditions is relevant for almost all the applications. Leaving to the numerical modeling the detailed treatment of specific cases, we analyse a few basic examples.

5.2.2.1 The Mean Shear Case

Let us consider a flow characterized by homogeneous and steady second-order moments (constant $\langle u_i u_j \rangle$) and a mean shear flow $U_1(x_3) = \Gamma x_3$: then $\tilde{\boldsymbol{u}} = (\Gamma x_3 + u_1, u_2, u_3) \equiv (\Gamma z + u, v, w)$. Being the initial position (the source) $\boldsymbol{x} = 0$, the second-order moments of the parcel positions $\langle x_i x_j \rangle_S$ can be computed from:

$$\langle x^2 \rangle_S = \int_0^t \mathrm{d}t_1 \int_0^t \mathrm{d}t_2 \langle [\Gamma z(t_1) + u(t_1)][\Gamma z(t_2) + u(t_2)] \rangle \ , \tag{5.15}$$

$$\langle xz \rangle_S = \int_0^t \mathrm{d}t_1 \int_0^t \mathrm{d}t_2 \langle [\Gamma z(t_1) + u(t_1)] w(t_2) \rangle \ , \tag{5.16}$$

$$\langle x_i^2 \rangle_S = \int_0^t \mathrm{d}t_1 \int_0^t \mathrm{d}t_2 \langle u_i(t_1) u_i(t_2) \rangle \ , \ i = 2, 3 \ . \tag{5.17}$$

For the exact solution see Monin and Yaglom (1971, p. 556) and Exercise 5.1.

The velocity gradient term introduces a time scale $T_s = |(\mathrm{d}U_1/\mathrm{d}x_3)^{-1}| = \Gamma^{-1}$, on which the inhomogeneity acts. For small times, i.e. $t \ll \min(T_{ii}, T_s)$ the ballistic behaviour is found. For large times, i.e. $t \gg T_s$:

$$\langle x^2 \rangle_S \simeq \frac{2}{3} \Gamma^2 \langle w^2 \rangle T_{33} t^3 \ , \tag{5.18}$$

$$\langle xz \rangle_S \simeq \Gamma \langle w^2 \rangle T_{33} t^2 \ , \tag{5.19}$$

while $\langle y^2 \rangle_S$ and $\langle z^2 \rangle_S$ grow as t. Thus, the inhomogeneity induces a dispersion process faster than the diffusive one, due to the different advection at the different positions in z. In the literature of statistical mechanics (see, for instance Bouchaud and Georges 1990) this is called 'superdiffusion'. Moreover, note that in the leading term of $\langle x^2 \rangle_S$ the coefficient is proportional to the z statistics: the variance of the vertical velocity and the integral time scale for the same component are driving the spread in the x direction.

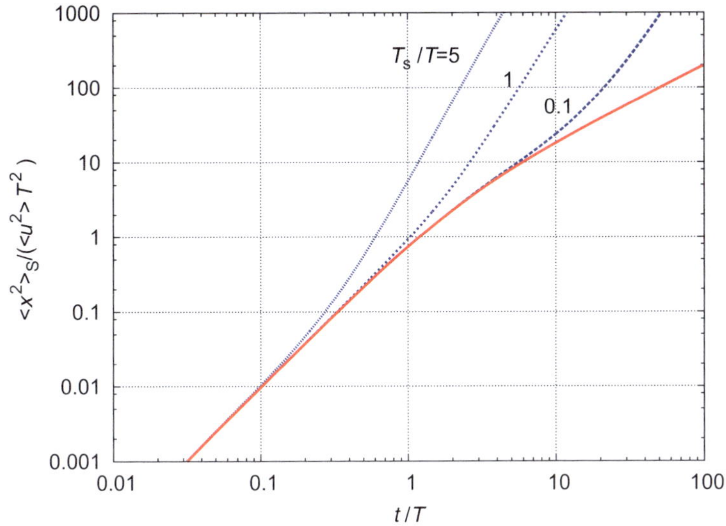

Fig. 5.1 Variance of the x coordinate of the fluid parcels normalized over the variance of the turbulent fluctuations and the square of the integral time scale, as a function of time normalized over the same scale. *Red line*: Eq. (5.11) with $R_{13} = 0$. *Blue lines*: solutions of (5.15) with $R_{13} = 0$, for $T_s/T = 0.1, 1, 5$

The solution is plotted in Fig. 5.1, using the exponential model (2.112) for the covariance and assuming $T_{11} = T_{33} = T$, $\langle u^2 \rangle = \langle w^2 \rangle$. If the shear time scale is smaller than the integral time scale (for the x direction), the diffusive behaviour cannot be seen. In the opposite case, the diffusive behaviour is detected in the interval $T < t < T_s$. This observation suggests a practical rule whether the shear effect can be neglect or not; in other words, to what extent in a inhomogeneous environment the diffusion paradigm can be used.

5.2.2.2 Effect of the Inhomogeneity of the Second-Order Moments on the Ballistic Regime

To analyze the ballistic behaviour in a gradient of velocity variance, let consider a steady velocity field with zero mean $\tilde{\boldsymbol{u}} = (u, v, w)$ and $\mathrm{d}\langle w^2 \rangle / \mathrm{d}z > 0$ (like in the CBL). In the other directions the second-order moments of the velocity do not change.

The vertical component of the velocity along a trajectory may be written in terms of the acceleration $\dot{w}$:

$$\frac{\mathrm{d}z}{\mathrm{d}t} \equiv w(t) = w(0) + \int_0^t \dot{w}(\tau)\,\mathrm{d}\tau \,, \tag{5.20}$$

where

$$\dot{w}(t) = \frac{\partial w}{\partial t} + u_j \frac{\partial w}{\partial x_j} . \tag{5.21}$$

Taking into account the continuity equation for the fluctuations (2.69), the time derivative of the average height of the trajectory ensemble results:

$$\frac{\mathrm{d}\langle z \rangle_S}{\mathrm{d}t} \simeq \frac{\mathrm{d}\langle w^2 \rangle}{\mathrm{d}z} t, \; t \ll T_{33} . \tag{5.22}$$

The inhomogeneity gives rise to a drift of the ensemble of trajectories, in spite of the fact that the mean velocity of the flow (the Eulerian velocity) is zero (Hunt 1985): the average is $\langle z \rangle = 0$, the conditional average is $\langle z \rangle_S \neq 0$.

5.2.2.3 A PBL Problem: The Effect of a Wall

It is interesting to look at the problem of dispersion in the PBL, in the light of the previous considerations, leaving aside a careful treatment based on proper modeling tools.

Based on dimensional arguments in neutral conditions, as discussed by Monin and Yaglom (1971, p. 563) or by Csanady (1973, p. 116), we argue that the average velocity components of the ensemble of parcels $\langle u_i \rangle_S$ must be only function of the PBL parameters u_*, $U(z)$ and of the time t. For symmetry reasons, $\langle u_2 \rangle_S = 0$. Thus:

$$\langle u_3 \rangle_S = b_3 u_*, \; \langle z \rangle_S = b_3 u_* t , \tag{5.23}$$

$$\langle u_1 \rangle_S = b_1 U(\langle z \rangle_S) \propto u_* \ln \frac{c u_* t}{z_0} . \tag{5.24}$$

Similarity arguments thus lead to argue that the average vertical position of the ensemble of parcels moves from the wall (the surface), because there is a physical limitation to the downwards displacements, but not to the upwards ones.

5.3 Two-Parcel Dispersion

The absolute dispersion deals with the statistics of independent parcels. The problem of the statistics of correlated parcels is tackled in its simpler way looking at the time evolution of the average separation of pair of parcels, released at the initial time at a distance (separation) smaller than the correlation scale of the turbulent flow (otherwise, they would be independent each other). Note that the pair dispersion has attracted much attention because of its relevance in the understanding of the turbulence, far beyond the problems we address in this book (Salazar and Collins

2009) and its extension to high-order correlations (such as tetrahedral shapes) is a subject of great interest.

5.3.1 The Dispersion in the Inertial Subrange

Let consider the separation $\boldsymbol{r}$ of a pair of parcels, whose trajectories are described by (2.1) with the condition (2.2), in steady homogeneous conditions. Thus:

$$\boldsymbol{r} \equiv \boldsymbol{x}^{(2)} - \boldsymbol{x}^{(1)} \\ = \Delta \boldsymbol{x}_0 + \int_0^t \boldsymbol{u}^{(2)}(\tau)\,\mathrm{d}\tau - \int_0^t \boldsymbol{u}^{(1)}(\tau)\,\mathrm{d}\tau = \boldsymbol{r}_0 + \int_0^t \Delta\boldsymbol{u}\,\mathrm{d}\tau\ , \tag{5.25}$$

where the superscript indicates the parcel.

The isotropy of the inertial subrange allows to neglect the direction of the separation vector, so we will consider the modulus of the velocity difference $\Delta u = |\Delta \boldsymbol{u}|$ and of the separation $r = |\boldsymbol{r}|$. The initial separation $r_0 = |\boldsymbol{r}_0|$ is larger than the Kolmogorov scale (2.92) and smaller than the integral length scale L_E. For short, r_0 is the size of the source.

Assume the flow homogeneous and steady: thus the variance of the velocity difference at a given time is related to the Eulerian structure function (2.95):

$$\langle \Delta u^2 \rangle = \langle |\boldsymbol{u}(\boldsymbol{x}^{(2)}) - \boldsymbol{u}(\boldsymbol{x}^{(1)})|^2 \rangle = 6[\langle u^2 \rangle - R(\langle r^2 \rangle^{1/2})]\ , \tag{5.26}$$

where the coefficient 6 accounts for the three components of the separation (it is 2 for each direction). In the inertial subrange the structure function grows as the separation to the $2/3$ power, so

$$\langle \Delta u^2 \rangle \propto \varepsilon^{2/3} \langle r^2 \rangle^{1/3}\ , \tag{5.27}$$

while when the separation is larger than the integral scale the covariance reduces to zero, and

$$\langle \Delta u^2 \rangle \propto \langle u^2 \rangle\ . \tag{5.28}$$

We study the variance of the separation

$$r = r_0 + \int_0^t \Delta u\,\mathrm{d}\tau \tag{5.29}$$

that is expected to depend on the value of the separation itself. Obviously, its mean value is $\langle r \rangle_\mathrm{S} = r_0$, where we consider the initial separation a deterministic variable.

Thus the variance of the stochastic variable r is:

$$\langle r^2\rangle_S = r_0^2 + \int_0^t dt_1 \int_0^t dt_2 \langle \Delta u(t_1)\Delta u(t_2)\rangle \ . \tag{5.30}$$

The ballistic regime occurs for $t < t_{r_0}$, where

$$t_{r_0} \approx r_0/\langle \Delta u^2\rangle(r_0)^{1/2} \approx \varepsilon^{-1/3} r_0^{2/3} \tag{5.31}$$

is the characteristic time of the eddies of the size of the source: the velocity difference is almost constant and the separation variance reads:

$$\langle r^2\rangle_S \approx r_0^2 + \langle \Delta u^2\rangle(r_0)\, t^2 \ , \tag{5.32}$$

where it is assumed that at time t the velocity difference is dominated by the eddies of scale r_0.

For longer times, when $r \gg r_0$, the separation involves eddies of larger and larger scales, and correspondingly larger and larger velocities. We assume that the leading term of the integrand in (5.30) is the velocity difference at time t; thus:

$$\langle r^2\rangle_S \approx r_0^2 + \langle \Delta u^2\rangle(\langle r^2\rangle_S^{1/2})\, t^2 \ . \tag{5.33}$$

Using (5.27), this expression leads to the Richardson (1926) law:

$$\langle r^2\rangle_S - r_0^2 = g\varepsilon t^3 \ , \tag{5.34}$$

that holds for $t_{l_0} \ll t \ll T$ and $l_0 \ll l \ll L_E$, namely, in the inertial subrange, where K41 is assumed to hold. Recent evaluations (Pagnini 2008; Sawford et al. 2008) suggest $g = 0.6$.

5.3.2 *The Diffusive Phase*

Let investigate the diffusive phase. From (5.25), we can explicitly consider the difference of the integrals and remember that, if the parcel separation is larger than L_E, the two-parcel velocity covariance becomes negligible. Thus the separation variance results from the sum of the variances of two independent parcels:

$$\langle r^2\rangle_S \approx r_0^2 + \left\langle \left(\int_0^t u^{(2)}\, d\tau \right)^2 \right\rangle + \left\langle \left(\int_0^t u^{(1)}\, d\tau \right)^2 \right\rangle , \tag{5.35}$$

which for $t \gg T$ is described by the diffusive phase of the absolute dispersion: for each direction the separation variance is twice the position variance of each parcel, and in our notation

$$\langle r^2\rangle_S \approx r_0^2 + 12\langle u^2\rangle Tt\,. \tag{5.36}$$

Note that the diffusive regime for the two-parcel dispersion problem is reached as both the conditions for long time and large distance are satisfied. If the Eulerian-to-Lagrangian time scale ratio (2.115) is of order one, these conditions are satisfied simultaneously.

5.4 Meandering

In the frame of absolute dispersion, when considering the analysis of experiments or of numerical simulations, it is relevant to distinguish single realizations (averaged over a set of independent particles) and the ensemble of realizations. This topic can be rephrased as follows: consider an experiment of absolute dispersion from a source, what can be understood from the analysis of many equivalent dispersion experiments? This problem is treated from a different point of view by Csanady (1973, pp. 94–96).

Let us consider a set of N particles, with coordinates $x_i^{(n)}(t)$, $i = 1, 2, 3$, $n = 1, \ldots, N$. Averaging over N is identified by $\langle\cdots\rangle$. (In this section, the conventions about the representation of averaging are autonomous from the rest of the book.)

The mean position is

$$\langle x_i\rangle = \frac{1}{N}\sum_{n=1}^{N} x_i^{(n)} \tag{5.37}$$

Define the displacement with respect to the mean position $y_i^{(n)} = x_i^{(n)} - \langle x_i\rangle$. Obviously, $\langle y_i\rangle = 0$.

The second order moments of the displacements are:

$$\langle y_i y_j\rangle = \langle x_i x_j\rangle - \langle x_i\rangle\langle x_j\rangle \tag{5.38}$$

Let us consider now M sets of N particles each. For each set out of M, moments are identified by $\langle\cdots\rangle^{(m)}$, with $m = 1, \ldots, M$. Averaging over M is identified by $\overline{\cdots}$.

The mean position is

$$\overline{\langle x_i\rangle} = \frac{1}{NM}\sum_{k=1}^{NM} x_i^{(k)} = \frac{1}{M}\sum_{m=1}^{M}\left(\frac{1}{N}\sum_{n=1}^{N} x_i^{(n+(m-1)N)}\right) = \frac{1}{M}\sum_{m=1}^{M}\langle x_i\rangle^{(m)} \tag{5.39}$$

The second-order moments are

$$\overline{\langle x_i x_j \rangle} = \overline{\langle y_i y_j \rangle} + \frac{1}{M} \sum_{m=1}^{M} \langle x_i \rangle^{(m)} \langle x_j \rangle^{(m)} = \overline{\langle y_i y_j \rangle} + \overline{\langle x_i \rangle \langle x_j \rangle} \tag{5.40}$$

which means that the covariance of the positions is made by the sum of the covariances of the displacements, plus the covariance of the mean positions (the meandering).

Let us consider the distance (the separation) between two particles, for each m set: $r_i = x_i^{(j)} - x_i^{(l)}$.

The variance of the distance is:

$$\langle r_i^2 \rangle \simeq \frac{1}{N^2} \sum_{j=1}^{N} \sum_{l=1}^{N} [x_i^{(j)} - x_i^{(l)}]^2 = 2\langle x_i^2 \rangle - 2\langle x_i \rangle^2 \tag{5.41}$$

Because of (5.38) it results

$$\langle r_i^2 \rangle \simeq 2\langle y_i^2 \rangle \tag{5.42}$$

namely, the variance of the distance between pair of particles is twice the variance of the displacements.

In the initial (ballistic) phase of dispersion, $\langle r_i^2 \rangle^{1/2} \ll L_{ii}$, so $\overline{\langle y_i^2 \rangle} \simeq 0$ and thus $\overline{\langle x_i^2 \rangle} \simeq \overline{\langle x_i \rangle^2}$. According to (5.7):

$$\overline{\langle x_i \rangle^2} = \langle u_i^2 \rangle t^2 \, , \tag{5.43}$$

i.e., the main contribution to the dispersion is due to the movement of center of mass of the puffs: there is sweeping but almost negligible growth due to eddies smaller than the puff itself.

For $t \gg T_{ii}$ and large separation, trajectories are independent each other. Thus, from (5.9):

$$\frac{\mathrm{d}\overline{\langle x_i^2 \rangle}}{\mathrm{d}t} = 2T_{ii} \langle u_i^2 \rangle \, , \tag{5.44}$$

and from (5.36), considering one component,

$$\frac{\mathrm{d}\overline{\langle y_i^2 \rangle}}{\mathrm{d}t} = 2T_{ii} \langle u_i^2 \rangle \, , \tag{5.45}$$

that is:

$$\frac{d\overline{\langle x_i \rangle^2}}{dt} = 0 , \tag{5.46}$$

i.e., the meandering is expected to become constant for long times.

5.5 Observations of Dispersion

In this section we discuss in detail the results of two basic experiments, one made in the laboratory, and the other in the field, to put into evidence some fundamental aspects discussed previously. Most of the atmospheric observations refer to more complex cases, which can be properly tackled using models, like those described in the following Sect. 5.6; details of their application are found in the literature.

5.5.1 Mikkelsen et al. (1987): Horizontal Meandering and Relative Dispersion in the Surface Layer

The experiment was made analyzing the smoke plume from a point source dispersing under relatively constant wind over a flat surface. Each release last for 40 min and was monitored up to a distance of about 1000 m. For each release, about 20 instantaneous observations were made. The full experiment is described by Mikkelsen (1983) and Mikkelsen and Eckman (1983).

The parameters we use to apply the Taylor (1921) formula (5.11) are: mean wind $U = 4.7\,\mathrm{m\,s^{-1}}$, lateral Eulerian velocity variance $\langle v^2 \rangle \simeq 1\,\mathrm{m^2\,s^{-2}}$, $T_\mathrm{E} \simeq 100\,\mathrm{s}$. Note that according to (2.115) the ratio between the Eulerian and the Lagrangian time scales is about 3, so we use $T_{22} = T_\mathrm{E}/3$ (at variance with the original paper).

In Fig. 5.2 the variances of the lateral displacements are reported: the absolute dispersion is obviously larger than the meandering and the relative dispersion: it results that the largest contribution to the absolute dispersion is given in this case by the dispersion of the center of mass, while the relative dispersion (the dispersion of the single puffs around their center of mass) is about one order of magnitude smaller.

5.5.2 Vertical and Lateral Dispersion in a Laboratory CBL

The experiment described by Snyder et al. (2002) and Weil et al. (2002) is made in a tank simulating the convective boundary layer and dispersion in it, revising

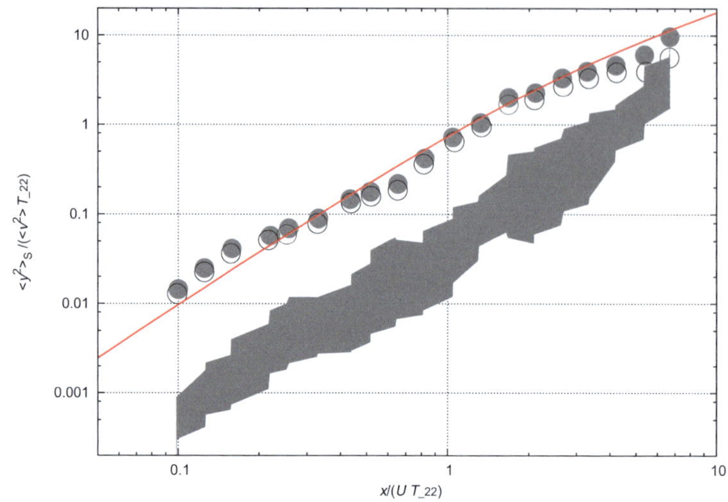

Fig. 5.2 Variances of the lateral displacements: data redrawn from Mikkelsen et al. (1987, Fig. 8). The variances are made nondimensional as in Fig. 5.1 and are reported as function of the downwind distance normalized over the advection distance covered by a parcel in the time interval corresponding to the Lagrangian time scale. *Full circles*: absolute dispersion; *open circles*: meandering; *gray area*: ensemble of observations of relative dispersion. The *red line* corresponds to (5.11) with the parameters given in the text

the classical measurements by Willis and Deardorff (1976). A number (33) of independent (with identical characteristics) releases was made, so the individual spread and the ensemble average can be evaluated. Some informations about the experiment: the standard deviations of the components of the velocity are estimated to be $\approx 0.6W_*$, the source height is $0.15h$. In showing the results, we take into account that the authors observed that the effective (in terms of dispersion observations) CBL depth was larger by 15 % of the density depth (as determined from the experimental setup). To compute the vertical drift of the mean position from (5.22), we have used a simplified form of (3.86), i.e. the normalized vertical velocity variance equal to $1.8(z/h)^{2/3}$.

The results are reported in Fig. 5.3. The lateral absolute dispersion is quite well described by (5.11). The meandering reaches a constant value for the lateral direction, in agreement with (5.46), and goes to zero for the vertical direction, because any single release has reached (at $t \approx h/W_*$) the well-mixed condition and all the centers of mass lie on the line $\langle z \rangle_S/h = 0.5$. (Thus for times larger than the integral time scale the relative dispersion and the absolute dispersion are almost superimposed.)

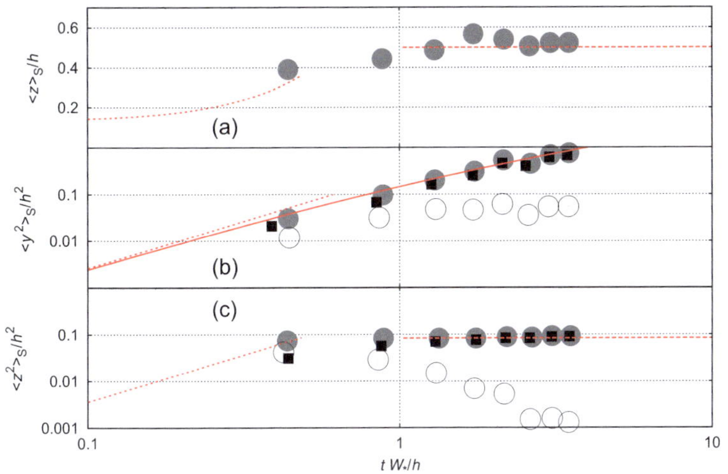

Fig. 5.3 Dispersion in the CBL, data redrawn from Weil et al. (2002). (**a**) Mean height (*full circles*), normalized over the CBL depth. (**b**) Lateral dispersion, normalized over the CBL depth squared. *Full circles*: absolute dispersion; *squares*: relative dispersion; *open circles*: meandering. (**c**) As in panel (**b**), for the vertical dispersion. In (**a**) the *dotted line* corresponds to (5.22) with (3.86) simplified as described in the text; the *dashed line* to the well mixed level $\langle z \rangle / h = 0.5$. In (**b**) the *dotted line* corresponds to the ballistic behaviour (5.43) and the continuous line to (5.11) with the parameters given in the text. In (**c**) the *dotted line* is as in (**b**) and the *dashed line* corresponds to the well mixed value $\langle z^2 \rangle / h^2 = 1/12$

5.6 The Stochastic Approach to the Absolute Dispersion of Tracer Parcels

In order to deal with non-ideal conditions it is necessary to develop proper modeling tools for the transport problem: either models for the trajectory of the tracer parcels in the phase space (traditionally indicated as Lagrangian ones) or models for the concentration field (Eulerian ones).

5.6.1 The Link Between the Eulerian and Lagrangian Descriptions

The link between the two approaches arises from a well known mathematical result (see, for an introductory discussion, Risken 1989, Sect. 1.2.3).

Let z the vector of dimensions N defining the state of the parcels: for instance, if we identify z with the parcel position then $N = 3$, whereas $N = 6$ is necessary if positions and velocities are used to define the state.

We assume that the evolution of the state from the initial condition $z_0 = z(t_0)$ to z is described by the Langevin equation (LE):

$$dz_i = a_i(z,t)\,dt + b_{ij}(z,t)\,dW_j(t),\ i,j = 1,\ldots,N\ , \tag{5.47}$$

where $dW_j(t)$ is a Wiener process such that

$$\langle\, dW_i(t)\rangle = 0\ , \tag{5.48}$$

and

$$\langle\, dW_i(t)\,dW_j(s)\rangle = \delta_{ij}\delta(t-s)\,ds\,dt\ . \tag{5.49}$$

The terms a_i and b_{ij} are called drift and diffusion term respectively.

The probability density function f of transition from z_0 to z satisfies the Fokker-Planck equation (FPE):

$$\frac{\partial f}{\partial t} = -\frac{\partial}{\partial z_i}(\alpha_i f) + \frac{\partial^2}{\partial z_i \partial z_j}\left(\frac{\beta_{ij}^2}{2} f\right)\ . \tag{5.50}$$

Note that the condition (5.49) may be written (Gardiner 1990, Sect. 4.2.6f):

$$\left\langle \int_0^t A(\tau)\,dW(\tau) \int_0^t B(\tau)\,dW(\tau)\right\rangle = \int_0^t \langle A(\tau)B(\tau)\rangle\ d\tau\ . \tag{5.51}$$

This result gives the link between the Lagrangian description of the evolution of the state of each parcel, Eq. (5.47), and the probability of finding parcels with state z at time t, given that the initial state is z_0 at t_0, Eq. (5.50). Because the probability density function of the positions is proportional to the mean concentration (5.2), this result gives the link with the diffusion equation.

We will refer to the models based on LE as Lagrangian Stochastic Models (LSM).

5.6.2 The Model with Uncorrelated Velocities: N = 3

To develop a model for the tracer parcel trajectories in the physical space, i.e. $z \equiv \boldsymbol{x}$, in the approximation of uncorrelated velocities, we will use (5.47) with $N = 3$:

$$dx_i = a_i\,dt + b_{ij}\,dW_j\ , \tag{5.52}$$

with initial conditions $x_i(0) = x_{i0}$.

Defining $D_{ij} = b_{ij}^2/2$, the corresponding FPE reads:

$$\frac{\partial f}{\partial t} = -\frac{\partial a_i f}{\partial x_i} + \frac{\partial^2}{\partial x_i \partial x_j}(D_{ij} f) \; . \tag{5.53}$$

The same equation holds for C, being proportional to f, and can be rewritten as:

$$\frac{\partial C}{\partial t} = -\frac{\partial}{\partial x_i}\left[\left(a_i - \frac{\partial D_{ij}}{\partial x_j}\right) C\right] + \frac{\partial}{\partial x_i}\left(D_{ij}\frac{\partial C}{\partial x_j}\right) \; . \tag{5.54}$$

This equation can be compared with the Reynolds averaged equation for C (2.66):

$$\frac{\partial C}{\partial t} = -\frac{\partial}{\partial x_i}\left(U_i C + \langle u_i c\rangle\right) + \kappa_c \frac{\partial^2 C}{\partial x_i \partial x_i} \; , \tag{5.55}$$

where the turbulent flux term $\langle u_j c\rangle$ appears.

Thus, the diffusion coefficient arises from a flux-gradient relationship:

$$-\langle u_i c\rangle + \kappa_c \frac{\partial C}{\partial x_i} = -D_{ij}\frac{\partial C}{\partial x_j} \tag{5.56}$$

and the drift velocity accounts for the mean flow velocity and the inhomogeneities of the diffusion coefficient:

$$a_i = U_i + \frac{\partial D_{ij}}{\partial x_j} \tag{5.57}$$

A similar identification is made by Thomson (1995). The same couple of Eqs. (5.52) and (5.54) is discussed by Mazzino et al. (2005, Eqs. (11) and (8)).

5.6.2.1 A Simple Example

In the simple case of homogeneous isotropic steady turbulence with no mean flow: $b_{ij} = \sqrt{2D}\delta_{ij}$, $U_i = 0$, and an instantaneous source in $\boldsymbol{x}_0 = 0$, the solution for the first and second moment of the position can be computed from (5.52): see Exercise 5.2 and Gardiner (1990, p. 102). It reads:

$$\langle x_i\rangle_S(t) = 0 \; , \tag{5.58}$$

$$\langle x_i^2\rangle_S(t) = 2Dt \; , \tag{5.59}$$

which is the same as the Taylor (1921) solution (5.9) for long times (the diffusive limit).

The diffusion equation (2.15) for an incompressible fluid flow applied to the mean concentration, with initial condition $C(\boldsymbol{x}, t) = \delta(\boldsymbol{x})\delta(t)$, no mean flow and isotropic constant diffusion coefficient D, has the well-know solution:

$$C(\boldsymbol{x}, t) = \frac{1}{(2\pi)^{3/2}(2Dt)^{3/2}} \exp\left[-\frac{x_i x_i}{4Dt}\right] , \tag{5.60}$$

from which the variance of the concentration, i.e., the variance of the position of the tracer parcels, results, for each component:

$$\langle x_i^2 \rangle_S(t) = \frac{\int_{-\infty}^{\infty} x_i^2 C(\boldsymbol{x}, t)\, \mathrm{d}x_i}{\int_{-\infty}^{\infty} C(\boldsymbol{x}, t)\, \mathrm{d}x_i} = 2Dt . \tag{5.61}$$

For the integrals, see Gradshteyn and Ryzhik (1980, p. 307 (3) and p. 338 (8)). This result highlights the equivalence of the Lagrangian (based on the Langevin equation) and Eulerian (based on the diffusion equation) descriptions.

5.6.2.2 The Extension to Compressible Flows

Because the model with $N = 3$ is widely used also for dispersion outside the PBL, i.e. it is the basis of the trajectory models for the whole atmosphere (Stohl et al. 2005; Draxler and Hess 1998), it is worth looking at its formulation in a variable density flow. The formulation was first suggested by Venkatram (1993) and then refined and generalised by Thomson (1995); see also Rossi and Maurizi (2014).

The WMC is rephrased stating that if $C \propto \langle \rho_f \rangle$ (the concentration is proportional to the ensemble average of the air density) at a time t_0, then for all $t > t_0$ the two quantities must remain proportional. This implies that $\langle \rho_f \rangle$ is also a solution of (5.53), or of (5.54).

Substituting f with $\langle \rho_f \rangle$ in (5.53) and averaging the continuity equation

$$\frac{\partial \langle \rho_f \rangle}{\partial t} = -\frac{\partial}{\partial x_i} \left(\overline{U}_i \langle \rho_f \rangle \right) , \tag{5.62}$$

where $\overline{U}_i$ is the density-weighted mean velocity (Thomson 1995), defined as:

$$\overline{U}_i = \frac{\langle \tilde{u}_i \rho_f \rangle}{\langle \rho_f \rangle} = U_i + \frac{\langle u_i \rho \rangle}{\langle \rho_f \rangle} , \tag{5.63}$$

the following expression is obtained:

$$-\frac{\partial}{\partial x_i} \left(\overline{U}_i \langle \rho_f \rangle \right) = -\frac{\partial}{\partial x_i} (a_i \langle \rho_f \rangle) + \frac{\partial^2}{\partial x_i \partial x_j} (D_{ij} \langle \rho_f \rangle) . \tag{5.64}$$

Then, integrating both sides and rearranging, it results:

$$a_i = \frac{\partial D_{ij}}{\partial x_j} + \frac{D_{ij}}{\langle \rho_f \rangle} \frac{\partial \langle \rho_f \rangle}{\partial x_j} + \overline{U}_i \,, \tag{5.65}$$

where the non-uniqueness in the integration is removed considering that in the well-mixed state, the mixing ratio flux is proportional to $\overline{U}_i \langle \rho_f \rangle$. Note that substituting (5.65) into (5.54) gives the equivalent of Eq. (2) in Thomson (1995).

The density variations imply an additional term in the drift: compare (5.57) and (5.65). Note that the density-weighted mean velocity is congruent with the concept of averaged velocity in a variable density flow: in the applications, $\overline{U}_i = U_i$.

5.6.3 *The Model with Uncorrelated Accelerations: N = 6*

The absolute dispersion in very high Reynolds number flow (so that the inertial subrange extends to very small spatial scales and viscosity is neglected) can be modeled for time scales both larger and smaller than the integral time scale (but larger than the time scale related to viscosity) using LE to model the velocity increments.

The fluid parcel trajectory is given by

$$\mathrm{d}x_i = \tilde{u}_i \, \mathrm{d}t \,. \tag{5.66}$$

The model for the parcel velocity $\tilde{\boldsymbol{u}}$ is:

$$\mathrm{d}\tilde{u}_i = a_i(\boldsymbol{x}, \tilde{\boldsymbol{u}}, t)\, \mathrm{d}t + b_{ij}(\boldsymbol{x}, t)\, \mathrm{d}W_j \,. \tag{5.67}$$

The challenge is the derivation of suitable forms for the terms a_i and b_{ij} of (5.67).

5.6.3.1 The Derivation of the LE Terms

The model formulation (Thomson 1987) must be made in such a way that the statistical properties (at least, some statistical properties) of the vector $z = (\boldsymbol{x}, \tilde{\boldsymbol{u}})$ are the same as for true fluid parcels, whose motion is exactly determined by the (unknown) solution of the Navier-Stokes equations. Of course, each realisation of the model does not correspond to an actual trajectory (note also that the trajectory estimated from the stochastic equation is continuous but not differentiable).

A couple of remarks about (5.67) has to be made.

To avoid ambiguities (see van Kampen 1981), we chose the tensor b_{ij} independent on $\tilde{\boldsymbol{u}}$. Secondly, while in general b_{ij} is non-diagonal, as in (5.47), in this treatment we have only one component of noise for each velocity component; this choice will be shown to be consistent with the small scale isotropy of K41.

The second-order moments of the velocity increments can be derived from (5.67):

$$\langle \mathrm{d}u_i \, \mathrm{d}u_j \rangle = b_{ik} b_{jl} \langle \mathrm{d}W_k \, \mathrm{d}W_l \rangle + O(\mathrm{d}t^{3/2}) \simeq b_{ik} b_{jl} \delta_{kl} \, \mathrm{d}t \equiv B_{ij} \, \mathrm{d}t \ , \tag{5.68}$$

for $\mathrm{d}t$ small with respect to the integral time scale (so that the increment of velocity refers only to the fluctuations) and to a time scale proper for the integration of the LE equation: $O(|\boldsymbol{u}|/|\boldsymbol{a}|)$.

For consistency with the inertial subrange relation for the Lagrangian velocity increments (2.111), it must be:

$$B_{ij} = C_0 \varepsilon \delta_{ij} \ , \tag{5.69}$$

so that we can define

$$B_{ij} = \tilde{B} \delta_{ij} \ , \tilde{B} = C_0 \varepsilon \ , \tag{5.70}$$

and the amplitude of the noise term in (5.67) can be defined consistently:

$$b_{ii} = \sqrt{C_0 \varepsilon} \ , \forall i \ . \tag{5.71}$$

Equations (5.66) and (5.67) describe the fluid parcel trajectories, from the initial point $(\boldsymbol{x}_0, \tilde{\boldsymbol{u}}_0)$ at time t_0 to $(\boldsymbol{x}, \tilde{\boldsymbol{u}})$ at time t, consistently with the presence of the inertial subrange.

The corresponding FPE (5.50) reads:

$$\frac{\partial f}{\partial t} = -\frac{\partial}{\partial x_i}(\tilde{u}_i f) - \frac{\partial}{\partial \tilde{u}_i}(a_i f) + \frac{C_0 \varepsilon}{2} \frac{\partial^2 f}{\partial \tilde{u}_i \partial \tilde{u}_i} \ . \tag{5.72}$$

Thomson (1987) applied to this equation the WMC (5.4): in a constant density fluid, if f is a solution of (5.72), also f_{E} must be a solution. If we assume that the probability distribution function of the Eulerian velocity f_{E} is known, (5.72) becomes an equation for a_i:

$$\frac{\partial}{\partial \tilde{u}_i}(a_i f_{\mathrm{E}}) = \frac{C_0 \varepsilon}{2} \frac{\partial^2 f_{\mathrm{E}}}{\partial \tilde{u}_i \partial \tilde{u}_i} - \frac{\partial f_{\mathrm{E}}}{\partial t} - \tilde{u}_i \frac{\partial f_{\mathrm{E}}}{\partial x_i} \ , \tag{5.73}$$

where the effects of the unsteadiness and of the inhomogeneity of the flow on the drift term are put into evidence by the presence of the derivatives of f_{E} with respect to time and position.

Equation (5.73) may be written as:

$$a_i f_E = \frac{C_0 \varepsilon}{2} \frac{\partial f_E}{\partial \tilde{u}_i} + \Phi_i(\boldsymbol{x}, \tilde{\boldsymbol{u}}, t) \,, \tag{5.74}$$

$$\frac{\partial \Phi_i}{\partial \tilde{u}_i} = -\frac{\partial f_E}{\partial t} - \frac{\partial \tilde{u}_i f_E}{\partial x_i} \,, \tag{5.75}$$

with the condition $|\Phi| \to 0$ as $|\tilde{\boldsymbol{u}}| \to \infty$. Note that any solenoidal field decaying fast enough can be added to Φ_i: i.e. the solution is not unique, in the general three-dimensional case (but not in the one-dimensional case). The non-uniqueness of the solution has been investigated for instance by Borgas and Sawford (1994) or by Reynolds (2002).

5.6.3.2 Discussion of the Model with Gaussian f_E

We will present and discuss an application of the model to a flow with Gaussian distribution for the velocity $\tilde{u}_i = U_i + u_i$ and covariance matrix $V_{ij}(\boldsymbol{x}, t) = \langle u_i u_j \rangle$:

$$f_E(\tilde{\boldsymbol{u}}, t) = \frac{1}{(2\pi)^{3/2} \det(V)^{1/2}} \exp\left[-\frac{1}{2} u_i (V^{-1})_{ij} u_j \right] . \tag{5.76}$$

A solution for a_i reads (Thomson 1987):

$$a_i = -\frac{C_0 \varepsilon}{2} (V^{-1})_{ik} u_k + \frac{\Phi_i}{f_E} \,, \tag{5.77}$$

with

$$\begin{aligned} \frac{\Phi_i}{f_E} &= \frac{1}{2} \frac{\partial V_{il}}{\partial x_l} + \frac{\partial U_i}{\partial t} + U_l \frac{\partial U_i}{\partial x_l} \\ &+ \left[\frac{1}{2} (V^{-1})_{lj} \left(\frac{\partial V_{il}}{\partial t} + U_m \frac{\partial V_{il}}{\partial x_m} \right) + \frac{\partial U_i}{\partial x_j} \right] u_j \\ &+ \frac{1}{2} (V^{-1})_{lj} \frac{\partial V_{il}}{\partial x_k} u_j u_k \,. \end{aligned} \tag{5.78}$$

Note that the covariance matrix in general presents non-zero off-diagonal terms (the momentum fluxes) that are related to the larger scales of the flow; K41 must be satisfied by the small-scale features. If the covariance is a diagonal tensor, the three-dimensional process reduces to three one-dimensional independent processes.

Let us consider here the case of the one-dimensional homogeneous process, with $U = 0$. Then f_E reads:

$$f_E(u,t) = \frac{1}{(2\pi)^{1/2}\langle u^2\rangle^{1/2}} \exp\left(-\frac{u^2}{2\langle u^2\rangle}\right) . \tag{5.79}$$

The drift term becomes:

$$a = -\frac{C_0\varepsilon}{2\langle u^2\rangle} u = -\frac{u}{T_{uu}} . \tag{5.80}$$

The full model reads:

$$dx = u\,dt , \tag{5.81}$$

$$du = -\frac{u}{T_{uu}}\,dt + (C_0\varepsilon)^{1/2}\,dW = -\frac{u}{T_{uu}}\,dt + \left(\frac{2\langle u^2\rangle}{T_{uu}}\right)^{1/2} dW . \tag{5.82}$$

The stochastic equation (5.82) can be integrated: see Gardiner (1990, p. 106) and Exercise 5.3. It results:

$$u(t) = u(0)\exp(-t/T_{uu}) + \left(\frac{2\langle u^2\rangle}{T_{uu}}\right)^{1/2} \int_0^t \exp[-(t-s)/T_{uu}]\,dW(s) . \tag{5.83}$$

The mean velocity of the parcels reads:

$$\langle u\rangle_S(t) = \langle u\rangle_S(0)\exp(-t/T_{uu}) , \tag{5.84}$$

where $\langle u\rangle_S(0)$ is the mean velocity of the marked parcels at the source.

The covariance function for the velocity fluctuations reads:

$$\langle u(t)u(s)\rangle_S = \left[\langle u(0)^2\rangle_S - \langle u^2\rangle\right]\exp[-(t+s)/T_{uu}] + \langle u^2\rangle\exp(-|t-s|/T_{uu}) . \tag{5.85}$$

Note that the covariance is a function of both the time t and of the delay $|t-s|$, because at $t = 0$ the velocity variance of the parcels is different from that of the flow. If the source is in equilibrium with the flow: i.e. $\langle u\rangle_S(0) = 0$, $\langle u(0)^2\rangle_S = \langle u^2\rangle$, the covariance is function of the delay only and reproduces the exponential covariance, as in (2.112).

Some statistics concerning the parcel position can be computed. The mean position is determined by integrating Eq. (5.83) and averaging:

$$\langle x\rangle_S(t) = \langle u\rangle_S(0)\,T_{uu}[1 - \exp(-t/T_{uu})] . \tag{5.86}$$

The variance of the position is computed from the definition (5.5) with $i = j = 1$ and reads:

$$\langle x^2\rangle_S(t) = T_{uu}^2(\langle u(0)^2\rangle_S - \langle u^2\rangle)[1 - \exp(-t/T_{uu})]^2$$
$$+2\langle u^2\rangle T_{uu}t - 2\langle u^2\rangle T_{uu}^2[1 - \exp(-t/T_{uu})] \,. \tag{5.87}$$

The ballistic $t \ll T_{uu}$ and diffusive $t \gg T_{uu}$ regimes can be computed for the variances of the velocity fluctuations and of the displacements $y = x - \langle x\rangle_S$:

$$\langle u^2\rangle_S(t) \simeq \langle u(0)^2\rangle_S \left(1 - 2\frac{t}{T_{uu}}\right) + 2\langle u^2\rangle \frac{t}{T_{uu}} \,, \tag{5.88}$$

$$\langle y^2\rangle_S(t) \simeq [\langle u(0)^2\rangle_S - \langle u\rangle_S(0)^2]t^2 - \left[\langle u(0)^2\rangle_S - \langle u\rangle_S(0)^2 - \frac{2}{3}\langle u^2\rangle\right]\frac{t^3}{T_{uu}} \,, \tag{5.89}$$

and

$$\langle u^2\rangle_S(t) \simeq \langle u^2\rangle \,, \tag{5.90}$$

$$\langle y^2\rangle_S(t) \simeq 2\langle u^2\rangle T_{uu}t \,. \tag{5.91}$$

Note that $\langle u(0)^2\rangle_S - \langle u\rangle_S(0)^2$ is the variance of the marked parcel velocity at the source.

In the ballistic regime the parcel statistics is related to the velocity distribution at the source, while the memory of the initial condition is lost in the diffusive regime, and the Eulerian statistics drives the process. The transition from the ballistic to the diffusive regime for the variance of the displacement occurs with a t^3 correction whose sign is positive or negative depending on the value of the initial velocity with respect to the Eulerian equilibrium velocity. If the source is in equilibrium with the flow, the coefficient of t^3 is given by $-C_0\varepsilon/6$.

This model is consistent with the basic Taylor (1921) formulation, and allows to take into account a statistics at the source different from that of the flow (for instance, a jet). More complex solutions allow to take into account for instance the presence of vorticity in the flow (see, for instance Sawford 1999; Reynolds 2006).

5.6.3.3 A Different Model Formulation

In geophysical applications, the stochastic model with $N = 6$ can be formulated assuming that the velocity field is made by a explicitly known (deterministic) part and by fluctuations. The known part can be the mean flow obtained from the observations, or the flow from a model, based on the RANS equations, or the resolved velocity field computed by a LES (see Chap. 6 for details). Note that the mean flow from models is intended as an ensemble average when RANS equations are considered (Reynolds stresses depend on the flow and not on the grid dimensions). On the other hand, in cases of instantaneous filtered flow field from LES, the fluctuations are a measure of the subgrid turbulence and depend on the grid size.

Here we reformulate the $N = 6$ model putting into evidence the known part of the velocity field and using the LSM for the fluctuations.

Let us consider the Eulerian probability density function (5.76) with the covariance matrix V_{ij} constant in time and space. The tracer parcel velocity is $\tilde{u}_i = U_i + u_i$. Its increment is:

$$\mathrm{d}\tilde{u}_i = \mathrm{d}U_i + \mathrm{d}u_i = \left[\frac{\partial U_i}{\partial t} + (U_k + u_k)\frac{\partial U_i}{\partial x_k}\right]\mathrm{d}t + \mathrm{d}u_i\,, \tag{5.92}$$

where the mean velocity variation along the tracer parcel trajectory has been accounted for.

Equation (5.66) reads:

$$\mathrm{d}x_i = \tilde{u}_i\,\mathrm{d}t = [U_i(\boldsymbol{x}, t) + u_i]\,\mathrm{d}t\,, \tag{5.93}$$

which states that the motion of the parcel has to be considered as the motion following the mean flow plus the fluctuation.

We shall formulate the stochastic model for the velocity, Eq. (5.67), in terms of the fluctuating components of the parcel velocity:

$$\mathrm{d}u_i = a_i\,\mathrm{d}t + (C_0\varepsilon)^{1/2}\,\mathrm{d}W_i \tag{5.94}$$

The (total) velocity increment (5.92) becomes:

$$\mathrm{d}\tilde{u}_i = \left[\frac{\partial U_i}{\partial t} + U_k\frac{\partial U_i}{\partial x_k} + u_k\frac{\partial U_i}{\partial x_k} + a_i\right]\mathrm{d}t + (C_0\varepsilon)^{1/2}\,\mathrm{d}W_i\,, \tag{5.95}$$

in which the first two terms on the RHS of (5.95) evidence the contribution from the variations of the mean flow and the third term describes the coupling between the fluctuating field and the inhomogeneous mean field. The inhomogeneity introduces the time scale $T_\mathrm{s} = |\partial U_i/\partial x_k|^{-1}$ (which is an extension of the one defined in Sect. 5.2.2.1), so this coupling may be neglected if $T_{ii} \ll T_\mathrm{s}$ and we are interested in times smaller than T_s.

Using (5.77) and (5.78) the term a_i can be computed, and results:

$$a_i = -\frac{u_i}{T_{ii}}\,, \tag{5.96}$$

according to (5.95).

Applications of this formalism into the 'wind plus turbulence' scheme of atmospheric motion is common. Applications to LES velocity fields are presented by Weil et al. (2004) or by Vinkovic et al. (2006b) for instance, with different choices of the parametrization of the sub-grid contribution.

5.6.3.4 Comments to the $N = 6$ Model

The stochastic model with $N = 6$ is the basis for many applicative models. Some comments are in order.

The use of the WMC to determine the drift term in LE implies the knowledge of the Eulerian probability distribution function of the velocity as a function of position and time. With explicit reference to the contents of Chaps. 3 and 4, we know at best some moments, measured in a few points, thus a model for the f_{E} must be adopted.

The Gaussian distribution can be used for the stable cases, and for the horizontal components of the velocity in any stability conditions (remember the discussion in 3.4.3.2). For the vertical component the CBL shows important departures from Gaussian shape. Note that CBL was the first case of application of LSM, because of the conceptual failure of the models based on the diffusion equation and of the complications necessary to explain the observations. For non-Gaussian distributions (which are expected to be frequent in geophysical flows), a general solution like (5.77) and (5.78) is not available. Some hint can be obtained looking at (5.73) in the one-dimensional, steady case. Note that in many practical applications, the cross-correlation is neglected, the three-dimensional dispersion problem reduces to three independent one-dimensional processes, so we can concentrate to the determination of the drift coefficient looking at the one-dimensional case.

The drift term (5.73) reads:

$$af_{\mathrm{E}} = \frac{C_0\varepsilon}{2}\frac{\mathrm{d}f_{\mathrm{E}}}{\mathrm{d}\tilde{u}} + \Phi\,, \quad \frac{\mathrm{d}\Phi}{\mathrm{d}\tilde{u}} = -\tilde{u}\frac{\mathrm{d}f_{\mathrm{E}}}{\mathrm{d}x} \tag{5.97}$$

and thus

$$\Phi = -\int_{-\infty}^{\tilde{u}} \tilde{u}'\frac{\mathrm{d}f_{\mathrm{E}}}{\mathrm{d}x}\,\mathrm{d}\tilde{u}' = -\frac{\mathrm{d}}{\mathrm{d}x}\int_{-\infty}^{\tilde{u}} \tilde{u}' f_{\mathrm{E}}\,\mathrm{d}\tilde{u}' \tag{5.98}$$

Probably one of the most used approaches, suggested by Baerentsen and Berkowicz (1984) and implemented according to the Thomson (1987) scheme by Luhar and Britter (1989), to represent the vertical velocity distribution in their horizontally-integrated model of dispersion in the CBL, is based on the superposition of two Gaussian distributions. This probability density function reads:

$$f_{\mathrm{E}}(\tilde{u},x) = \frac{a_1}{\sqrt{2\pi}\sigma_1}\exp\left[-\frac{(\tilde{u}-m_1)^2}{2\sigma_1^2}\right] + \frac{a_2}{\sqrt{2\pi}\sigma_2}\exp\left[-\frac{(\tilde{u}-m_2)^2}{2\sigma_2^2}\right], \tag{5.99}$$

and its 6 coefficients (a_i, σ_i, m_i) are determined in order to satisfy the known moments of the considered component. Free parameters (four conditions result from the imposition of the moments up to the third one) are used to avoid undesired features (see, for instance Maurizi and Tampieri 1999).

Another choice is the Maximum Missing Information probability density function (Jaynes 1957):

$$f_{\mathrm{E}}(\tilde{u}, x) = \exp\left(-\sum_{k=0}^{M} \lambda_k(x)\tilde{u}^k\right) , \tag{5.100}$$

with M even and the coefficients λ_k obtained from the normalization condition and from the knowledge of the first M moment. For $M = 2$ this probability density function reduces to a Gaussian distribution. Using (5.100) a polynomial expression for the drift term in the homogeneous non-Gaussian case, from (5.97), is obtained:

$$a = -\frac{C_0\varepsilon}{2}\sum_{k=1}^{M} k\lambda_k\tilde{u}^{k-1} , \tag{5.101}$$

(the highest order power of the velocity is odd).

A further possibility is based on the moment approximation, exploited by Kaplan and Dinar (1993) and used for instance in the one-dimensional, steady model of Franzese et al. (1999). Accordingly, the drift term is expressed as a second-order polynomial in the velocity $a(\tilde{u}, x) = \alpha(x)\tilde{u}^2 + \beta(x)\tilde{u} + \gamma(x)$. From the FPE (5.73) the expressions for the first three moments of $\tilde{u}$ can be derived as functions of the parameters α, β and γ, which are then determined equating the expressions to the measured (or estimated) moments. This assumption has been shown to work approximately in a limited velocity range, so its applicability is problem-dependent.

A further important point is that the stochastic model based on (5.67) with (5.68) is strictly related to the existence of an inertial sublayer, which allows to identify the diffusion term in the FPE. Note that to estimate the dissipation ε from (2.113) (Tennekes 1982) it should be accounted for that the Lagrangian decorrelation time scale is not equal in general to the Lagrangian integral time scale if the distribution function of the velocity is non-Gaussian (Maurizi and Lorenzani 2001).

To model dispersion in some geophysical conditions (not uncommon in the ocean, but also in low wind conditions in the atmosphere, for instance), the observations suggest to split the flow field into two components with different correlation times, and then to model the flow field by the sum of two stochastic processes (independent each other). The correct definition of the drift and diffusion terms needs special care: see for instance Maryon (1998) or Pasquero et al. (2001) for atmospheric and oceanic applications respectively.

5.7 Dispersion of Inertial Particles

Inertial particles, like aerosols and dust, are important components of the atmosphere; their dispersion is relevant in most environmental problems, so that the departure of their behaviour from that of the fluid parcels is worth to be discussed.

Finite-size particles with density different from the fluid are subject to many forces (Maxey and Riley 1983; Maxey 1987): here we concentrate on particles of size much smaller than the Kolmogorov scale of the flow (point particles) and of density much greater than that of the fluid. Such particles move because of the gravity field $\boldsymbol{g} = (0, 0, -g)$ and of the friction due to the viscous fluid: their velocity is different from that of the flow and the condition (2.2) is replaced by the drag law:

$$\frac{\mathrm{d}v_i}{\mathrm{d}t} = -\delta_{i3}g + \frac{\tilde{u}_i[\boldsymbol{x}(t)] - v_i}{T_\mathrm{p}} , \tag{5.102}$$

where the particle position is given by (2.1) and the particle relaxation time (the Stokes time) T_p is:

$$T_\mathrm{p} = \frac{m_\mathrm{p}}{6\pi\rho_\mathrm{f}\nu a} = \frac{2}{9}\frac{\rho_\mathrm{p}}{\rho_\mathrm{f}}\frac{a^2}{\nu} . \tag{5.103}$$

Here $m_\mathrm{p} = (4/3)\pi a^3 \rho_\mathrm{p}$ is the particle mass, a its radius, ρ_p its density.

Equation (5.102) with (5.103) holds in the continuous regime (particle size much larger than the molecular free path) and for low particle Reynolds number $|\tilde{\boldsymbol{u}} - \boldsymbol{v}|a/\nu < 1$.

The solution of (5.102) for $t/T_\mathrm{p} \to \infty$ in a still fluid gives the terminal velocity of the particle:

$$v_\mathrm{s} = T_\mathrm{p} g . \tag{5.104}$$

The general solution of (5.102) in the absence of gravity and for zero mean flow velocity ($\tilde{u}_i = u_i$, $U_i = 0$) reads (see Exercise 5.4):

$$v_i(t) = \frac{1}{T_\mathrm{p}} \int_{-\infty}^{t} \mathrm{e}^{(s-t)/T_\mathrm{p}}\, u_i(s)\, \mathrm{d}s , \tag{5.105}$$

and the corresponding velocity variance is (see Exercise 5.5):

$$\langle v_i^2 \rangle = \frac{\langle u_i^2 \rangle}{1 + T_\mathrm{p}/T} , \tag{5.106}$$

being T the Lagrangian time scale.

Csanady (1963) pointed out that the inertial particles move differently from the fluid because of the "crossing trajectories" effect (because their velocity is different from that of the flow, the Lagrangian flow velocity structure of the particle is different from that of a fluid parcel; for large drift the Lagrangian structure becomes similar to the Eulerian structure, sampled at almost fixed time, for separation given by the elapsed time and the particle actual drift velocity), and because of the "inertia" effect (which means that there is a delay for the particle to follow the flow: they respond to accelerations more slowly than fluid parcels). More recently, theoretical analyses and DNS have shown that the heavy particles are expelled from rotating regions (high vorticity regions, non-hyperbolic regions), leading to clustering (Cencini et al. 2006; Bec et al. 2007): the particle velocity field is compressible even in an incompressible flow. The Csanady (1963) approach neglects clustering.

To understand and model the particle dispersion in a turbulent flow, T_p and v_s must be compared with the proper characteristic time and velocity scales of the flow: the Kolmogorov scale $\mathrm{T_K}$, which is the shortest time scale of the turbulence, the integral time scale T_E and the vertical velocity scale $\langle w^2\rangle^{1/2}$.

For $T_p \ll \mathrm{T_K}$ the viscous drag is large enough to ensure that the particle follows the turbulent velocity field (i.e., the response time is short enough to allow them to follow the flow evolution): for the Stokes number $S_t \equiv T_p/\mathrm{T_K} \rightarrow 0$ the particles behave approximately as the fluid parcels. As the Stokes number increases but still is less than unity, particles are able to follow the motion, but are ejected from the rotating regions so that the acceleration standard deviation of the particles decreases and clustering in hyperbolic regions of the flow occurs. The standard deviation of the particle acceleration is reported in Fig. 5.4a, from the numerical simulations by Cencini et al. (2006). This ejection process (the clustering) is more evident at $S_t \approx 0.5$, then decreases, as shown in Fig. 5.4b, because particles decorrelate from the flow structure. For similar values of S_t, the standard deviation of the fluid parcel acceleration conditioned to be in the particle position has a minimum. As S_t increases the inertia acts as a filter: the highest frequencies of the spectrum become uneffective, and for $T_p \gg T_E$ ideally all turbulence is filtered. It results that the standard deviation of the acceleration of the fluid particles, filtered along the particle trajectory over a time window of the size of the Stokes time, is larger than that of the particles at small S_t, and converges to it at large S_t (see Fig. 5.4a, blue and red lines).

With reference to the turbulence in the quasi-neutral surface layer, the situation is visualized in Fig. 5.5, considering a ratio of particle to air density equal to 1000, and a friction velocity range $0.1\,\mathrm{m\,s^{-1}} < u_* < 1\,\mathrm{m\,s^{-1}}$. We use $T_E = \kappa z/u_*$ and the dissipation rate $\varepsilon = u_*^3/(\kappa z)$, necessary to compute $\mathrm{T_K}$. It results that all the particles with radius less than about 100 μm are expected to follow the fluctuations of the flow velocity at $z = 1$ m (radius below the blue dashed line, corresponding to $S_t = 1$), while the maximum dimension reduces of a factor 3 considering $z = 0.01$ m, an height comparable with the grass roughness length. At the same time, these particles are subject to clustering, so the variance of their velocity is expected to be smaller than that of the flow, more than that predicted by (5.106). The line corresponding to $S_t = 5$ puts into evidence the conditions for which clustering is negligible: particles of larger radius are expected to have velocity variance described by (5.106).

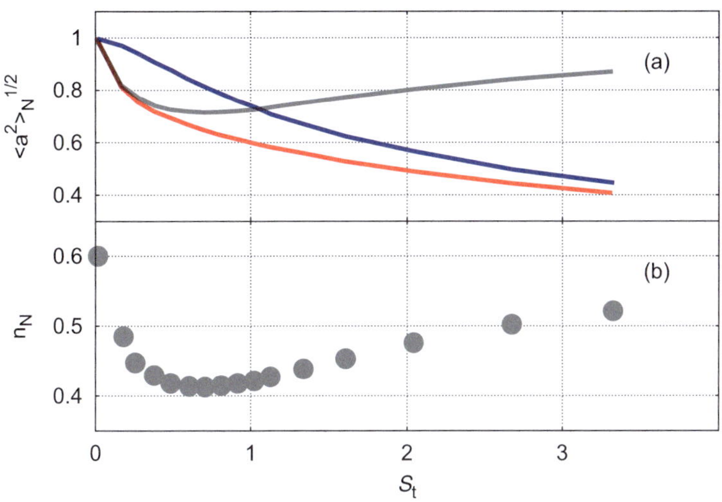

Fig. 5.4 (**a**) Standard deviation of the acceleration of inertial particles (*red line*), of the flow parcels sampled in the positions of the inertial particles (*gray line*) and of the flow parcel filtered along the particle trajectory over a time window of the size of the Stokes time, according to Eq. (5.105) (*blue line*) normalized by the flow acceleration, as function of the Stokes number of the particles, in stationary conditions. (**b**) Number density of inertial particles, normalized by the total density, in the non-hyperbolic regions of the flow. Results from DNS, redrawn from Cencini et al. (2006)

Thus, inertia can be neglected for large Stokes numbers (which broadly guarantees also that the Stokes time is much smaller than the Eulerian time scale), for instance, in the applications related to the dispersion in the main body of the flow, but it is expected to be important for the motion near solid surfaces (the deposition of particles is different from that of fluid parcels, i.e. of a passive tracer).

The relative dispersion is obviously affected by the Stokes number (Bec et al. 2010): broadly speaking, the Richardson t^3 regime is preceded by a ballistic-like regime (separation growing as t^2) related to the different velocities of the particles and of the flow, that lasts for a time related to the initial separation and the Stokes number itself.

The effect of gravity can be accounted for in a simplified way, considering that the settling velocity of particles of radius larger than about 100 μm is larger than the vertical velocity scale (radius above the black line in Fig. 5.5): the settling velocity has the effect to drift the particle across the turbulent eddies. In the limit of $v_s \gg \langle w^2 \rangle^{1/2}$, the Lagrangian (along a trajectory) covariance of the particle velocity is equal to the Eulerian covariance of the flow velocity for a separation $T_p^2 g$, which is the space covered by the parcel during a time interval of the order of the relaxation time. If such distance is larger than the integral scale of the eddies $L_E = T_E \langle w^2 \rangle^{1/2}$ the solid particle sees an almost uncorrelated velocity field (radius above the dashed-dotted lines in Fig. 5.5).

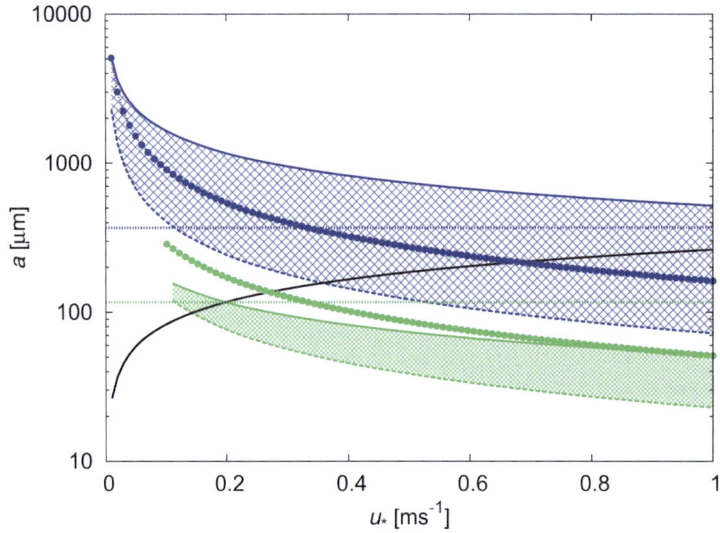

Fig. 5.5 Contour lines of the ratios $T_p/T_E = 1$ (*continuous*), $S_t = T_p/T_K = 1$ (*dashed*) and $S_t = 5$ (*dashed with circles*) as function of the friction velocity and of the particle radius, for neutral surface layer conditions. *Blue*: $z = 1$ m, *green*: $z = 0.01$ m. The *shaded areas* broadly correspond to the inertial sublayer. The *black line* corresponds to $v_s = u_*$. The *dashed-dotted horizontal lines* correspond to the radius of particles for which $v_s T_p = L_E$: *blue* and *green* refer to the heights as before

In Fig. 5.6 the wind tunnel observations by Snyder and Lumley (1971) are reported and compared with (5.106). The relevant parameters are reported in Table 5.1. From Fig. 5.6a it results that the velocity variance is overestimated by the simple model (5.106), because it do not take into account the clustering effect.

5.7.1 *The Parameterization of the Integral Time Scales for Particles*

Based on the previous considerations, Sawford and Guest (1991) formulated a parameterization to use LSM for heavy particle dispersion modeling, reviewed by Reynolds (2000).

Let assume that the mean particle velocity is the terminal velocity.

Then three cases are considered:

- for vanishing inertia and vanishing terminal velocity, the particle is equivalent to a fluid parcel (case 1);

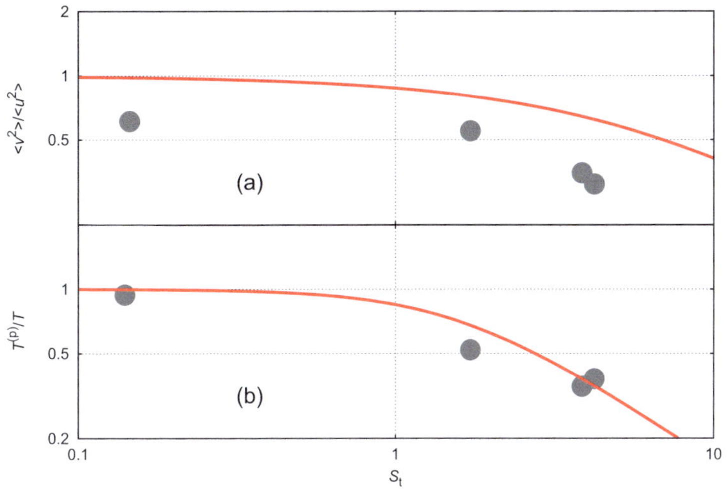

Fig. 5.6 (**a**) Particle velocity variance normalized over the flow velocity variance as function of the Stokes number. *Dots*: observations from Snyder and Lumley (1971). *Continuous line*: Eq. (5.106). (**b**) Particle Lagrangian time scale normalized over the flow time scale. *Dots* from Snyder and Lumley (1971). *Continuous line*: Eq. (5.113)

Table 5.1 Parameters from Snyder and Lumley (1971)

	S_t	v_s (m s^{-1})	$T^{(p)}/T$
Hollow glass	0.145	1.7×10^{-2}	0.94
Corn	1.72	0.20	0.52
Solid glass	3.85	0.44	0.35
Copper	4.21	0.48	0.38

Flow data: $\langle v^2 \rangle^{1/2} = 0.13\,\mathrm{m\,s^{-1}}$, $T_K = 0.012\,\mathrm{s}$, $T = 0.085\,\mathrm{s}$

- in the limit of large terminal velocity the particle trajectory approaches a straight line through the fluid and the fluid correlation along the trajectory approximates the Eulerian correlation (with a space separation equal to $v_s t$). In this case, the gravity imposes a specific direction affecting statistics: longitudinal (in the direction of $\boldsymbol{g}$) different from lateral (case 2);
- in the limit of large inertia ($T_p \gg T_E$) and small terminal velocity ($v_s \ll \langle w^2 \rangle^{1/2}$) the particle is stationary and the correlation reduces to Eulerian fixed point correlation (case 3).

The model for the velocity fluctuations in homogeneous and steady conditions, with null mean flow velocity, reads:

$$\mathrm{d}x_i = v_i\,\mathrm{d}t\,, \tag{5.107}$$

$$\mathrm{d}v_i = \left(\frac{u_i(\boldsymbol{x}(t), t) - v_i}{T_p} - g\delta_{i3}\right)\mathrm{d}t\,, \tag{5.108}$$

$$du_i = a_i\, dt + \sqrt{2\langle u_i^2\rangle / T_i^{(p)}}\, dW_i\,, \tag{5.109}$$

$$\frac{\partial}{\partial u_i}(a_i f_E) = \frac{\langle u_i^2\rangle}{T_i^{(p)}}\frac{\partial^2 f_E}{\partial u_i \partial u_i} - \frac{\partial f_E}{\partial t} - u_i\frac{\partial f_E}{\partial x_i}\,. \tag{5.110}$$

Note that (5.109) and (5.110) differ from (5.74) and (5.73) respectively because the effective integral time scale for the particle along a trajectory (the Lagrangian time scale for particle) $T^{(p)}$ is used. $T^{(p)}$ is obtained interpolating between the integral time scale for fluid elements T_E and the time scale L_E/v_s appropriate for large terminal velocities (case 1 to case 2). Note that L_E takes different values for different velocity components: in particular the length scale parallel to the separation vector (and thus parallel to $\boldsymbol{g}$) is twice the perpendicular one. The interpolation formulas are (Sawford and Guest 1991; Reynolds 2000):

$$T_{\parallel}^{(p)} = T\left[1 + \left(\frac{\beta v_s}{\langle w^2\rangle^{1/2}}\right)^2\right]^{-1/2} = T\left[1 + \left(\frac{T\, v_s}{L_E}\right)^2\right]^{-1/2}, \tag{5.111}$$

$$T_{\perp}^{(p)} = T\left[1 + \left(\frac{2\beta v_s}{\langle w^2\rangle^{1/2}}\right)^2\right]^{-1/2}, \tag{5.112}$$

where $\beta = T/T_E \approx 1/3$ is the inverse of the Eulerian-to-Lagrangian time scale ratio (2.115). Note that (5.112) for instance can be written as:

$$T_{\perp}^{(p)} = T\left[1 + \left(\frac{2\beta g T_K}{\langle w^2\rangle^{1/2}} S_t\right)^2\right]^{-1/2}. \tag{5.113}$$

Note that the case 3 is recovered approximately because $T^{(p)} \simeq T_E$ and being T_p large the effect of viscous drag is negligible, so that $dv_i \simeq 0$ from (5.108). From this model it results that the Lagrangian time scale for the particles is always shorter than the time scale for the fluid parcels. A test can be made against the data by Snyder and Lumley (1971), who give the values of $T^{(p)}$ derived from the observed correlation functions for the particles: see Fig. 5.6b.

This model has been considered for aerosol dispersion modelling (Wilson 2000; Vinkovic et al. 2006a), and within the accuracy of the data, it is shown that the inertial particle behaviour differs from that of fluid parcels mainly very near the surface, as deposition is concerned.

Exercises

5.1 Derive the explicit form of the integrals (5.15)–(5.17).

5.2 Integrate the stochastic equation (5.52) for the $N = 3$ model.

5.3 Integrate the stochastic equation (5.82) for the $N = 6$ model.

5.4 Find the solution of (5.102) with $g = 0$.

5.5 Derive the spectra and the variances for the inertial particles according to (5.102).

References

J.H. Baerentsen, R. Berkowicz, Monte-carlo simulation of plume diffusion in the convective boundary layer. Atmos. Environ. **18**, 701–712 (1984)

J. Bec, L. Biferale, M. Cencini, A.S. Lanotte, S. Musacchio, F. Toschi, Heavy particle concentration in turbulence at dissipative and inertial scales. Phys. Rev. Lett. **98**, 084502 (2007)

J. Bec, L. Biferale, A.S. Lanotte, A. Scagliarini, F. Toschi, Turbulent pair dispersion of inertial particles. J. Fluid Mech. **645**, 497–528 (2010)

P.S. Berloff, J.C. McWilliams, Material transport in oceanic gyres. Part II: hierarchy of stochastic models. J. Phys. Oceanogr. **32**, 797–830 (2002)

M.S. Borgas, B.L. Sawford, A family of stochastic models for two-particle dispersion in isotropic homogeneous stationary turbulence. J. Fluid Mech. **279**, 69–99 (1994)

J.-P. Bouchaud, A. Georges, Anomalous diffusion in disordered media: statistical mechanisms, models and physical applications. Phys. Rep. **195**, 127–293 (1990)

M. Cencini, J. Bec, L. Biferale, G. Boffetta, A. Celani, A.S. Lanotte, S. Musacchio, F. Toschi , Dynamics and statistics of heavy particles in turbulent flows. J. Turbul. **7**(36), 1–17 (2006)

G.T. Csanady, Turbulent diffusion of heavy particles in the atmosphere. J. Atmos. Sci. **20**, 201–208 (1963)

G.T. Csanady, *Turbulent Diffusion in the Environment* (D. Reidel Publishing Company, Dordrecht, 1973)

R.R. Draxler, G.D. Hess: An overview of the hysplit_4 modelling system for trajectories, dispersion, and deposition. Aust. Meteorol. Mag. **47**, 295–308 (1998)

P. Franzese, A.K. Luhar, M.S. Borgas, An efficient Lagrangian stochastc model of vertical dispersion in the convective boundary layer. Atmos. Environ. **33**, 2337–2345 (1999)

C.W. Gardiner, *Handbook of Stochastic Methods for Physics, Chemistry and the Natural Sciences*, 2nd ed. (Springer, Berlin, 1990)

I.S. Gradshteyn, I.M. Ryzhik, *Table of Integrals, Series, and Products*, 1980th ed. (Academic Press, New York, 1980)

J.C.R. Hunt, Turbulent diffusion from sources in complex flows. Ann. Rev. Fluid Mech. **17**, 447–485 (1985)

M.B. Isichenko, Percolation, statistical topography, and transport in random media. Rev. Mod. Phys. **64**, 961–1043 (1992)

E.T. Jaynes, Information theory and statistical mechanics. Phys. Rev. **106**, 620–630 (1957)

H. Kaplan, N. Dinar, A three-dimensional model for calculating the concentration distribution in inhomogeneous turbulence. Bound.-Layer Meteorol. **62**, 217–245 (1993)

P.E. Kloeden, E. Platen, *Numerical Solution of Stochastic Differential Equations*. Applications of Mathematics, vol. 23 (Springer, Berlin, Heidelberg, 1992)

J. Lighthill, *Fourier Analysis and Generalised Functions. Students' Edition* (Cambridge University Press, Cambridge, 1958)
A.K. Luhar, R.E. Britter, A random walk model for dispersion in inhomogeneous turbulence in a convective boundary layer. Atmos. Environ. **23**, 1911–1924 (1989)
R.H. Maryon, Determining cross-wind variance for low frequency wind meander. Atmos. Environ. **32**, 115–121 (1998)
A. Maurizi, S. Lorenzani, Lagrangian time scales in inhomogeneous non-Gaussian turbulence. Flow Turbul. Combust. **67**, 205–216 (2001)
A. Maurizi, F. Tampieri, Velocity probability density functions in Lagrangian dispersion models for inhomogeneous turbulence. Atmos. Environ. **33**, 281–289 (1999)
M. Maxey, The gravitational settling of aerosol particles in homogeneous turbulence and random flow fields. J. Fluid Mech. **174**, 441–465 (1987)
M.R. Maxey, J.J. Riley : Equation of motion for a small rigid sphere in a nonuniform flow. Phys. Fluids **26**, 883–889 (1983)
A. Mazzino, S. Musacchio, A. Vulpiani, Multiple-scale analysis and renormalization for preasymptotic scalar transport. Phys. Rev. E **71**, 011113.1–011113.11 (2005)
T. Mikkelsen, The borris field experiment: observations of smoke diffusion in the surface layer over homogeneous terrain (1983)
T. Mikkelsen, R. Eckman, Instantaneous observations of plume dispersion in the surface layer, in *14 th International Technical Meeting on Air Pollution Modelling and Its Applications*, Denmark, Copenhagen, 27–30 Sep 1983. ed. by C.D. Wispelaere
T. Mikkelsen, S.E. Larsen, H.L. Pecseli, Diffusion of Gaussian puffs. Quart. J. R. Meteor. Soc. **113**, 81–105 (1987)
A.S. Monin, A.M. Yaglom, *Statistical Fluid Mechanics*, vol. I, 769 pp. (MIT Press, Cambridge, 1971)
A.S. Monin, A.M. Yaglom, *Statistical Fluid Mechanics*, vol. II, 874 pp. (MIT Press, Cambridge, 1975)
E.A. Novikov, The Lagrangian-Eulerian probability relations and the random force method for nonhomogeneous turbulence. Phys. Fluids **29**, 3907–3909 (1986)
G. Pagnini, Lagrangian stochastic models for turbulent relative dispersion based on particle pair rotation. J. Fluid Mech. **616**, 357–395 (2008)
C. Pasquero, A. Provenzale, A. Babiano, Parameterization of dispersion in two-dimensional turbulence. J. Fluid Mech. **439**, 279–303 (2001)
A.M. Reynolds, On the formulation of lagrangian stochastic models for heavy-particle trajectories. J. Colloid Interface Sci. **232**, 260–268 (2000)
A.M. Reynolds, On the dynamical content of Lagrangian stochastic models in the Well-Mixed class. Bound.-Layer Meteorol. **103**, 143–162 (2002)
A.M. Reynolds, On chirality and turbulent dispersion. Physica D **218**, 185–190 (2006)
L.F. Richardson, Atmospheric diffusion shown on a distance-neighbor graph. Proc. R. Soc. Lond. Ser. A **110**, 709–737 (1926)
H. Risken, *The Fokker-Planck Equation. Methods of Solution and Applications*, 2nd ed. (Springer, Berlin, 1989)
D. Rossi, A. Maurizi, IL-GLOBO (1.0) – integrated lagrangian particle model and Eulerian general circulation model globo: development of the vertical diffusion module. Geosci. Model Dev. **7** 2181–2191 (2014)
J.P. Salazar, L.R. Collins, Two-particle dispersion in isotropic turbulent flows. Ann. Rev. Fluid Mech. **41**, 405–432 (2009)
B.L. Sawford, Rotation of trajectories in Lagrangian stochastic models of turbulent dispersion. Bound.-Layer Meteorol. **93**, 411–424 (1999)
B.L. Sawford, P.K. Yeung, J.F. Hackl, Reynolds number dependence of relative dispersion statistics in isotropic turbulence. Phys. Fluids (1994-present) **20**(6), 065111 (2008)
B.L. Sawford, F.M. Guest, Lagrangian statistical simulation of the turbulent motion of heavy particles. Bound.-Layer Meteorol. **54**, 147–166 (1991)

W.H. Snyder, J.L. Lumley, Some measurements of particle velocity autocorrelation function in a turbulent flow. J. Fluid Mech. **48**, 41–71 (1971)

W.H. Snyder, R.E. Lawson Jr., M.S. Shipman, J. Lu, Fluid modelling of atmospheric dispersion in the convective boundary layer. Bound.-Layer Meteorol. **102**, 335–366 (2002)

A. Stohl, C. Forster, A. Frank, P. Seibert, G. Wotawa, Technical note: the Lagrangian particle dispersion model FLEXPART version 6.2. Atmos. Chem. Phys. Discuss. **5**, 2461–2474 (2005)

G.I. Taylor, Diffusion by continuous movements. Proc. Lond. Math. Soc. **20**, 196–211(1921)

H. Tennekes, Similarity relations, scaling laws and spectral dynamics, in *Atmospheric Turbulence and Air Pollution Modeling*, pp. 37–68 (Reidel, Dordrecht, 1982), ed. by F.T.M. Nieuwstadt, H. van Dop

D.J. Thomson, Criteria for the selection of stochastic models of particle trajectories in turbulent flows. J. Fluid Mech. **180**, 529–556 (1987)

D.J. Thomson, Discussion. Atmos. Environ. **29**, 1343 (1995)

N.G. van Kampen, *Stochastic Processes in Physics and Chemistry* (North-Holland, Amsterdam, 1981)

A. Venkatram, The parameterization of the vertical dispersion of a scalar in the atmospheric boundary layer. Atmos. Environ. **27A**, 1963–1966 (1993)

I. Vinkovic, C. Aguirre, M. Ayrault, S. Simoens, Large-eddy simulation of the dispersion of solid particles in a turbulent boundary layer. Bound.-Layer Meteorol. **121**, 283–311 (2006a)

I. Vinkovic, C. Aguirre, S. Simoens, Large-eddy similation and lagrangian stochastic modeling of passive scalar dispersion in a turbulent boundary layer. J. Turbul. **7**, 1–14 (2006b)

J.C. Weil, W.H. Snyder, R.E. Lawson Jr., M.S. Shipman, Experiments on buoyant plume dispersion in a laboratory convection tank. Bound.-Layer Meteorol. **102**, 367–414 (2002)

J.C. Weil, P.P. Sullivan, C.-H. Moeng, The use of large-eddy simulations in lagragian particle dispersion models. J. Atmos. Sci. **61**, 2877–2887 (2004)

G. Willis, J.W. Deardorff, A laboratory model of diffusion into the convective planetary boundary layer. Q. J. R. Meteorol. Soc. **102**, 427–445 (1976)

J.D. Wilson, Trajectory models for heavy particles in atmospheric turbulence: comparison with observations. J. Appl. Meteorol. **39**, 1894–1912 (2000)

Chapter 6
Numerical Modeling of Turbulence for PBL Flows

Abstract Some general ideas about the turbulence representation in numerical models of PBL flows are presented. The parameterizations (closures) are discussed with reference to RANS and LES models. Complex flows can be described only via numerical simulations: some broad fields of application are exemplified.

6.1 Introduction

In Chaps. 3–5 the phenomenology of the PBL has been examined and interpreted on the bases of the physics of turbulence and of simplified models. From the basic equations of a stratified fluid we have elaborated a quantitative picture of the phenomena characterizing the turbulence in the PBL and its effects. This approach allows us to understand and describe the general (universal) features of the flow, but is unable to deal with the complex evolution of the real world and the many interactions among the various components of the system (the atmospheric system and its boundaries). To reach this objective, numerical models are necessary, which can give solutions unattainable with a direct analytical approach. As discussed in the Introduction of the book, the numerical models are also key tools to realize experiments (and to obtain data) for conditions difficult to select or to evidence from field experiments.

To numerically model turbulent flows, the options are:

- DNS (Direct Numerical Simulations): numerical solutions of the relevant fluid equations, for instance, the Navier-Stokes equations (2.3), the continuity equation (2.10) and the temperature equation (2.25);
- LES (Large Eddy Simulations): from the NSE and the other relevant equations, by taking the volume average, the resulting equations describe explicitly, broadly speaking, the scales larger than the scale characterizing the averaging volume, while the smaller ones must be parametrized;
- numerical solutions of the RANS equations, and/or of the equations of some ensemble averaged higher-order moments, which will supply informations on the probability density function of the variables: this approach, widely used in the previous chapters, requires a closure for the moments of order $n + 1$ that enter in the equations of the moments of order n;

F. Tampieri, *Turbulence and Dispersion in the Planetary Boundary Layer*,
Physics of Earth and Space Environments, DOI 10.1007/978-3-319-43604-3_6

- probability density function methods, that solve the equations for the probability density function of the variables of interest, formally derived from the NSE. We will not deal with these methods.

A detailed introduction to the different approaches can be found in Pope (2000).

The problem common to all the approaches but DNS is the need of parameterizations (closures), in order to substitute the quantities not explicitly treated by the equations to be solved with estimates of such quantities derived from the solution itself. In the next sections, the theoretical grounds of some exemplar parameterizations are presented.

6.2 Closures for the Reynolds-Averaged Equations

Dealing with the equations for the mean quantities (velocity, temperature, concentration), we need an expression for the second-order moments in terms of the first-order ones. The basic assumption to close the equations lies in the formulation of a flux-gradient relationship (Boussinesq 1877). The use of the flux-gradient relationship implies the existence of a diffusive process, or at least that the turbulent exchanges can be approximated by such a process. As already noted, this implies that the integral scales are smaller than the scales of the problem under examination, otherwise further approximations are necessary.

6.2.1 The Eddy Diffusion Model for the RANS Equations

The flux-gradient relationship for the Reynolds stresses reads:

$$\langle u_i u_j \rangle = \frac{1}{3}\langle u_i u_i \rangle \delta_{ij} - K_\tau \left(\frac{\partial U_i}{\partial x_j} + \frac{\partial U_j}{\partial x_i} \right) , \tag{6.1}$$

and the analogous relation for any scalar (here, written for the temperature) reads:

$$\langle u_i \theta \rangle = -K_Q \frac{\partial \Theta}{\partial x_i} . \tag{6.2}$$

According to Pope (2000, p. 359), the model (6.1) is based on the hypothesis that the Reynolds stress anisotropy is an effect of the mean gradients, which is true in shear-dominated turbulence.

Note that here we consider the eddy diffusion coefficients K_τ and K_Q as scalars, but this can be a crude simplification. For instance, in Sect. 3.4.3.4 it has been observed that the horizontal heat flux occurs also without any horizontal temperature gradient (because the eddies have a structure, at variance with the molecular free flights) and it was empirically related to the vertical gradients of mean quantities. Thus to cope with this phenomenon, and the like ones, (6.2) has to be written using a tensor form for the eddy diffusion coefficient, with at least the term $K_{Q13} \neq 0$ (Monin and Yaglom 1971, Sect. 10.3).

6.2.1.1 Eddy Diffusion Coefficient Based on the Mean Shear

In a fundamental paper about numerical simulation of atmospheric flows, Smagorinsky (1963) suggests a generalization of the expression (2.84):

$$K_\tau = 2l_\tau^2 \left(S_{ij}S_{ij}\right)^{1/2} , \tag{6.3}$$

where the mean rate-of-strain tensor is:

$$S_{ij} = \frac{1}{2}\left(\frac{\partial U_i}{\partial x_j} + \frac{\partial U_j}{\partial x_i}\right) , \tag{6.4}$$

and the problem is to give an expression for the mixing length for momentum l_τ. Note that (6.3) reduces to (2.84) for horizontally homogeneous conditions.

The definition of the mixing length is obviously quite empirical. It is important to note that with the present approach we imagine that the entire spectrum of turbulent eddies contributes to the flux, so that the mixing length is related to the integral length scale. We already found that in the logarithmic layer $l_\tau = \kappa z$. A practical formulation for the mixing length (in neutral conditions) was given by Blackadar (1962):

$$l_\tau^{-1} = (\kappa z)^{-1} + l_0^{-1} , \tag{6.5}$$

where a maximum mixing length is prescribed: $l_0 \sim 100$ m is a common choice. A number of definitions is reported for instance by Cuxart et al. (2006, Table III).

The eddy diffusion coefficient for heat is expected to be different from that of momentum, although the simple model of Sect. 2.4.6 says that they are equal, and both depend on stability (Sects. 3.4.2.3 and 3.7.1.3).

6.2.1.2 A Clue Concerning Heterogeneity

In the horizontally homogeneous case, (6.3) with (6.4) reduces to:

$$K_\tau^{(H)} = l_\tau^2 \left|\frac{\mathrm{d}U}{\mathrm{d}z}\right| . \tag{6.6}$$

Imagine a change in the x direction, similar to the heterogeneities discussed in Chap. 4, Sect. 4.2 or 4.4, leading to a variation of U with x, and neglect all other changes. It results:

$$K_\tau \approx l_\tau^2 \left[4 \left(\frac{\partial U}{\partial x} \right)^2 + \left(\frac{\partial U}{\partial z} \right)^2 \right]^{1/2}$$

$$= K_\tau^{(H)} \left[1 + 4 \frac{(\partial U/\partial x)^2}{(\partial U/\partial z)^2} \right]^{1/2} , \tag{6.7}$$

which shows that, in this formulation, the eddy diffusion coefficient is enhanced by the occurrence of heterogeneous conditions, with respect to the horizontally homogeneous case, as observed for instance in the discussion about the blending height (Sect. 4.4.1).

6.2.1.3 Eddy Diffusion Coefficient Based on TKE

The expression (6.3) prescribes no mixing in the absence of shear, while even in simple homogeneous conditions we know that turbulence produces mixing, as shown by (5.6) for instance. The expression (5.6) suggests also that the eddy diffusion coefficient depends on the variance of the velocity, so that, in isotropic approximation, it can be assumed to be proportional to the TKE, according to Kolmogorov (1942) and Prandtl (1945) (see Pope 2000, p. 369). Consistently with (5.10) and (2.115), the eddy diffusion coefficient can be expressed as:

$$K_\tau = c E_\mathrm{k}^{1/2} l_\tau , \tag{6.8}$$

and requires an estimate for the mixing length. Note that in the quasi-neutral surface layer this expression gives the same result of (6.3), by choosing a suitable value for c, which can be determined using the neutral values for the normalized variances of Table 3.3: $c \approx 0.4 \div 0.5$.

To use (6.8) for modelling purposes, it is necessary to solve (2.73), together with the RANS equations: to close it, an expression for the third-order moments is necessary:

$$-\left(\frac{\langle p u_j \rangle}{\rho_{00}} + \frac{1}{2} \langle u_i u_i u_j \rangle \right) = K_\tau \frac{\partial E_\mathrm{k}}{\partial x_j} , \tag{6.9}$$

where the same eddy diffusion coefficient is used. An expression for the dissipation rate can be derived from its equilibrium value:

$$\varepsilon = C_\varepsilon \frac{E_\mathrm{k}^{3/2}}{l_\tau} , \tag{6.10}$$

where, using the above-cited surface layer estimates for variances in quasi-neutral flow, it results $C_\varepsilon \approx 0.05 \div 0.14$ (Pope (2000, p. 372) suggests $C_\varepsilon \approx 0.17$).

The TKE equation then reads:

$$\frac{\mathrm{D}E_\mathrm{k}}{\mathrm{D}t} = \mathscr{P} + \mathscr{B} + \frac{\partial}{\partial z}\left(K_\tau \frac{\partial E_\mathrm{k}}{\partial z}\right) - \varepsilon \tag{6.11}$$

where the parameterization for the third-order moments (6.9) and the equilibrium expression for dissipation (6.10) are used; the shear production $\mathscr{P}$ and the buoyancy $\mathscr{B}$ terms are defined by Eqs. (3.6) and (3.7).

This approach is the basis of the Mellor-Yamada closure (Mellor and Yamada 1974), widely used in the meteorological applications.

Instead of using (6.10), the dissipation rate can be derived by integration of the semiempirical equation (Pope 2000, Eqs. (10.53) and (10.54)):

$$\frac{\mathrm{D}\varepsilon}{\mathrm{D}t} = -C_1 \langle u_i u_j \rangle \frac{\partial U_i}{\partial x_j}\frac{\varepsilon}{E_\mathrm{k}} - C_2 \frac{\varepsilon^2}{E_\mathrm{k}} + \frac{\partial}{\partial x_j}\left(\frac{K_\tau}{C_3}\frac{\partial \varepsilon}{\partial x_j}\right), \tag{6.12}$$

where suggested values are $C_1 = 1.44$, $C_2 = 1.92$ and $C_3 = 1.3$. The use of the prognostic equations for E_k and ε leads, from (6.8) and (6.10), to an expression for the eddy diffusion coefficient which do not involve the a-priori specification of the mixing length:

$$K_\tau = c_1 \frac{E_\mathrm{k}^2}{\varepsilon}, \tag{6.13}$$

with $c_1 = 0.09$ (Pope 2000, Eq. (10.47)).

As discussed at the end of the previous section, in presence of non-neutral conditions a stability correction is necessary.

6.2.2 The Closure for the Second-Order Moment Equations

An improvement of the flow description can be obtained by the explicit solution of the second-order moment equations, with an appropriate closure for the third-order moments, based on flux-gradient relationships between third-order moments and the gradients of the second-order ones, like (6.9).

Looking at the equations for the second-order moments for the velocity components (2.70), the temperature variance (2.75) and the heat fluxes (2.78), it results that the closure will require expressions for $\langle u_i u_j u_k \rangle$, for $\langle u_i p \rangle$, for $\langle u_i u_j \theta \rangle$, for $\langle u_i \theta^2 \rangle$ and for the respective dissipation rates. A classical review is presented by Mellor and Yamada (1982); the notation here is slightly different: see the note at the end of Exercise 6.1. The scheme presented here outlines the basic approach: modified (improved) parameterizations can be found in the literature and are continuously developed under the pressure of improving weather and climate forecasts.

6.2.2.1 The Third-Order Term Closure

The terms containing the pressure can be closed as follows (Rotta 1951):

$$\frac{1}{\rho_{00}}\left\langle p\left(\frac{\partial u_i}{\partial x_j}+\frac{\partial u_j}{\partial x_i}\right)\right\rangle = -\frac{E_{\mathrm{k}}^{1/2}}{3l_1}\left(\langle u_i u_j\rangle - \frac{2\delta_{ij}}{3}E_{\mathrm{k}}\right) + C_1 E_{\mathrm{k}}\left(\frac{\partial U_i}{\partial x_j}+\frac{\partial U_j}{\partial x_i}\right), \tag{6.14}$$

$$\frac{1}{\rho_{00}}\left\langle p\frac{\partial \theta}{\partial x_j}\right\rangle = -\frac{E_{\mathrm{k}}^{1/2}}{3l_2}\langle u_j\theta\rangle\ , \tag{6.15}$$

where l_1 and l_2 are assumed to be proportional to the mixing length; $C_1 \approx 0.16$ from Mellor and Yamada (1982). According to this parameterization, an anisotropic flow tends to isotropy in absence of external forcings.

The terms containing third-order covariances of velocity components and temperature are expressed as follows:

$$\langle u_k u_i u_j\rangle = -E_{\mathrm{k}}^{1/2}\lambda_1\left(\frac{\partial\langle u_i u_j\rangle}{\partial x_k}+\frac{\partial\langle u_i u_k\rangle}{\partial x_j}+\frac{\partial\langle u_j u_k\rangle}{\partial x_i}\right) \tag{6.16}$$

$$\langle u_k u_j\theta\rangle = -E_{\mathrm{k}}^{1/2}\lambda_2\left(\frac{\partial\langle u_k\theta\rangle}{\partial x_j}+\frac{\partial\langle u_j\theta\rangle}{\partial x_k}\right) \tag{6.17}$$

$$\langle u_k\theta^2\rangle = -E_{\mathrm{k}}^{1/2}\lambda_3\frac{\partial\langle\theta^2\rangle}{\partial x_k} \tag{6.18}$$

where λ_i are assumed to be proportional to the mixing length.

6.2.2.2 The Dissipation Terms and the Small Scale Isotropy

According to K41 (and remember Sect. 2.5.3), we assume the existence of an inertial sublayer characterized by isotropic conditions and high Reynolds number. So, from (2.72):

$$\varepsilon_{ik} \equiv 2\nu\left\langle\frac{\partial u_i}{\partial x_j}\frac{\partial u_k}{\partial x_j}\right\rangle = \frac{2}{3}\delta_{ik}\varepsilon = \frac{2}{3}\delta_{ik}\frac{E_{\mathrm{k}}^{3/2}}{\Lambda_1}\ , \tag{6.19}$$

from (2.79):

$$(\nu+\kappa_{\mathrm{H}})\left\langle\frac{\partial u_i}{\partial x_j}\frac{\partial\theta}{\partial x_j}\right\rangle = 0\ , \tag{6.20}$$

and from (2.77):

$$\varepsilon_\theta = \kappa_\mathrm{H} \left\langle \left(\frac{\partial \theta}{\partial x_j} \right)^2 \right\rangle = \frac{E_\mathrm{k}^{1/2} \langle \theta^2 \rangle}{\Lambda_2} , \tag{6.21}$$

and, as before, Λ_i are assumed to be proportional to the mixing length. Note that (6.19) is consistent with (6.10).

6.2.2.3 The Boundary Layer Approximation

The paradigmatic conditions discussed in Chap. 3, plus the steadiness, are often used in the meteorological practice to simplify the closures.

The so-called boundary layer approximation (Mellor and Yamada 1982, level 2.5) consists in the following assumptions:

- horizontal homogeneity;
- prognostic equation for TKE;
- diagnostic equations for other moments;

The TKE equation is given by (6.11) with the dissipation from (6.19), while a set of algebraic relationships are used for the second-order moments but TKE.

The eddy diffusion coefficients have the general form (6.8), but the quantity c is now different as far as momentum and heat are concerned, and is function of the stability (expressed in terms of the gradient Richardson number or the Obukhov length). Thus

$$K_\tau = c_\tau(R_\mathrm{g}) \sqrt{E_\mathrm{k}} l_\tau , \tag{6.22}$$

$$K_\mathrm{Q} = c_\mathrm{Q}(R_\mathrm{g}) \sqrt{E_\mathrm{k}} l_\tau , \tag{6.23}$$

The full set of the algebraic relationships for the moments is reported by Mellor and Yamada (1982) and in Exercise 6.1. As an example, note the relation for the horizontal heat flux:

$$\langle u\theta \rangle = \frac{3 l_2}{E_\mathrm{k}^{1/2}} \left[-\langle uw \rangle \frac{\partial \Theta}{\partial z} - \langle w\theta \rangle \frac{\partial U}{\partial z} \right] \tag{6.24}$$

which reduces to the heuristic expression (2.86) by proper substitutions.

6.2.2.4 The Lower Boundary Condition

Assume that the vertical discretization of the model is such that the lowest resolved level of integration lies within the surface layer, so that the surface layer similarity functions (Chap. 3) can be used as boundary conditions for the diagnostic equations: namely, for the momentum, temperature and TKE.

6.2.2.5 Stability Corrections

The stability effects on the eddy diffusion coefficients have been discussed in Sects. 3.4.2.3 and 3.7.1.3, where also the turbulent Prandtl number dependence on stability was shown: see Fig. 3.28c.

As an example from SBL, in Fig. 6.1 the eddy diffusion coefficients for momentum and heat, derived applying the flux-gradient relationship to the SABLES98 data at the lower level ($z = 5.8$ m), divided by the square root of the TKE measured at the same level, are reported and compared with the relationships derived by Zilitinkevich et al. (2013, Eq. (95)) and the parameterization used in a NWP model (Buzzi et al. 1994, 2003; Zampieri et al. 2005):

$$c_\tau = \sqrt{0.17}\, l_\tau\, (1 + 12R_g)^{-1}\,, \tag{6.25}$$

$$c_Q = c_\tau (1 + 5R_g)^{-1}\,, \tag{6.26}$$

where the mixing length is defined as in (6.5) with $l_0 = 100$ m.

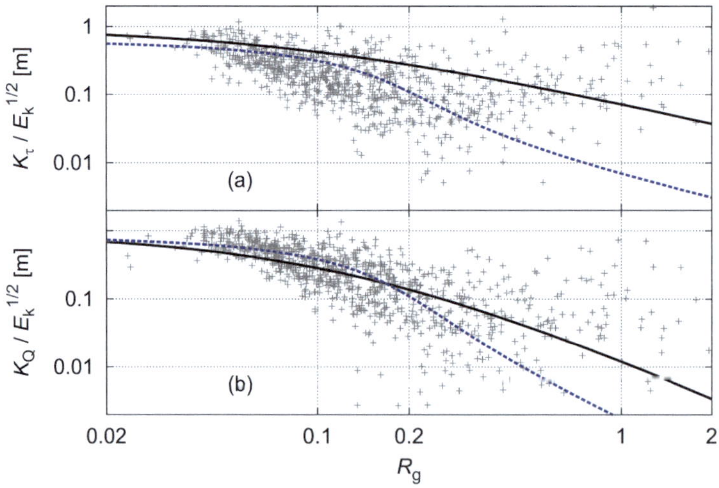

Fig. 6.1 Eddy diffusion coefficients for momentum (**a**) and heat (**b**) over the square root of the TKE, from SABLES98 data at $z = 5.8$ m (*crosses*). *Black line*: parameterization from the NWP model by Buzzi et al. (1994, 2003); Zampieri et al. (2005). *Blue line*: relationships from Zilitinkevich et al. (2013, Eq. (95)), with a modified numerical coefficient to fit the data

6.2.3 TKE and TPE Based Models

According to the arguments of Sect. 3.7, in stable conditions the TKE evolution equation (2.73) cannot describe the entire phenomenology occurring in strong stable conditions, and more generally if the turbulence is not mainly driven by the shear production. Thus, an evolution of the models based on the TKE closure consists in models which take into account also the evolution of the mean turbulent potential energy TPE, Eq. (3.13).

A complete scheme is presented by Zilitinkevich et al. (2013). The general outline is the same as before: the number of prognostic equations for the second-order moments is reduced solving only the evolution equations for TKE and TPE, and using steady state approximations for the other quantities.

A relevant feature is that the difference between the dissipation time scales of the TPE (i.e., the dissipation time scale of the temperature variance) and of the TKE is explicitly treated in this formulation. In fact, from the surface layer approximation, the ratio E_k/E_P can be evaluated as function of the flux Richardson number, with the dissipation rates expressed as $\varepsilon = E_k/t_\tau$ and $\varepsilon_P = E_P/(c_P t_\tau)$ as in Zilitinkevich et al. (2013, Eqs. (19), (53), (54)), and can be estimated from the experimental data, as shown in Fig. 3.30c. Note that $\varepsilon_P \propto \varepsilon_\theta$, see (3.13).

This modelling approach can be seen as an extension towards modelling non-K41 flows.

Note that a simplified version consists in the use of an equation for the mean total turbulent energy (given by the sum of TKE and TPE): see Mauritsen et al. (2007).

6.2.4 The CBL and the Problem of Non-diffusive Behaviour (Counter-Gradient Fluxes)

Any closure based only on the local properties cannot work in the CBL, where the scale of the energy-containing eddies is of the order of the depth of the boundary layer. Thus, the vertical fluxes of momentum and of scalars are not consistent with a flux-gradient relationship with a finite positive diffusion coefficient (see Sawford and Guest 1987, Fig. 7c).

As far closures of the first order equations are concerned, the simplest approach is to split the vertical flux of a scalar quantity $A + a$ into a local (diffusive) term and a non-local term, prescribed from a parameterization derived from the data and/or based on models of up- and down-draughts. Thus:

$$\langle aw \rangle = -K_a \frac{dA}{dz} + \langle aw \rangle_{NL} , \tag{6.27}$$

(see, for instance Deardorff 1966; Wyngaard 1987; Holtslag and Moeng 1991; Brown and Grant 1997).

Usually, K_a is taken to be proportional to the product of a velocity and a time scale, as in (6.3) or in (6.8). The velocity scale is a function of height, as suggested by the relation (3.86), and the length scale is related to the boundary layer depth. For instance, Brown and Grant (1997) suggest for momentum:

$$K_\tau = w_\tau \kappa h(z/h)(1 - z/h)^2 , \tag{6.28}$$

where $w_\tau = u_*/\Phi_U$ for $z < 0.1h$ and $w_\tau = (u_*^3 + 0.6W_*^3)^{1/3}$ for $z > 0.1h$: it is evident the need to cope with a range of unstable conditions, as well as with surface layer properties. Different authors suggest different parameterizations.

Any parameterization of the non-local term must account for the fact that this flux goes to zero approaching neutral conditions; as suggested by Wyngaard (1987), the flux of a scalar may be described as the combination of the surface (upwards) flux and the entrainment (downwards) flux. Using expressions similar to the one discussed for the heat flux (3.97), the non-local term can be related to the surface properties. As an example, from Brown et al. (2008), neglecting density variations with height, we can write for the non-local part of the momentum flux:

$$\langle u_i w \rangle_{NL} = \frac{CW_*^3}{w_\tau^3}\left[\frac{z'}{h'}\left(1 - \frac{z'}{h'}\right)^2\right]\langle u_i w \rangle_0 \ \text{ for } 0.1h < z < 0.9h , \tag{6.29}$$

where $z' = z - 0.1h$ and $h' = 0.9h$ and $C = 2.7$.

The heat flux has been examined by many authors. As an example, Holtslag and Moeng (1991) discussed a formulation based on (6.27) and obtained:

$$\langle w\theta \rangle = -K_Q \frac{\mathrm{d}\Theta}{\mathrm{d}z} + C \frac{\langle w^3 \rangle}{\langle w^2 \rangle^{3/2}} \langle w\theta \rangle_0 , \tag{6.30}$$

where $K_Q \propto \langle w^2 \rangle h/W_*$. This formulation stresses the importance of the asymmetry of the vertical velocity probability distribution function, which is related to the non-locality of the transport.

Closures for the second-order moments require a parameterization of the third-order ones, and the diffusive approach outlined in Sects. 6.2.2.1 and 6.2.2.3 is not satisfactory. For the same reasons that lead to the second term on the RHS of (6.30), the vertical fluxes of second-order moments can be expressed as function of the skewness of the vertical velocity:

$$\langle w^2 a \rangle = \frac{\langle w^3 \rangle}{\langle w^2 \rangle} \langle wa \rangle , \tag{6.31}$$

as done by Zilitinkevich et al. (1999) for the heat flux, and widely discussed by Mironov (2009, Eqs. (19) and (20), for instance).

Models based on higher-order closures are also in use, but are not treated here. They originated to model convection (Canuto 1992) and extended to SBL (see, for instance Ferrero et al. 2011). Out of curiosity, it must be noted that a standard approach to model fourth-order moments is to apply the quasi-Normal hypothesis (roughly, the probability density function of the variable has the same kurtosis of a Gaussian one), which was forwarded for the first time by Millionshchikov (1941). See for a detailed discussion Monin and Yaglom (1975, p. 241) and an example of application in Gryanik and Hartmann (2002).

6.3 Large Eddy Simulations

In the Large Eddy Simulation (LES), the unsteady turbulent motions of the larger (and energetic) eddies are directly represented, whereas the effects of smaller scale motions are modelled. This approach was motivated formerly by the meteorological applications, and later has been adopted in a wider range of applications. A comprehensive treatment of the LES approach is given for instance in Sagaut (1998), while a discussion focussed of the PBL is found in Wyngaard (2010).

The application of a filtering procedure separates the smaller scales of the motion from the larger ones. The most obvious choice is to identify the smaller scales as belonging to the inertial subrange: in terms of Fourier components, the filter must act at a spatial scale smaller than the integral Eulerian scale and larger than the Kolmogorov scale. Assuming the universality of the inertial subrange, this choice allows a simple parameterization of the scales smaller than the filter scale.

The straightforward approach is the use of a spectral filter, i.e. to work in the spectral space (with spectral models); however the common practice in PBL applications is to use the physical space: taking an average over a box makes simpler the interpretation of the results, although it is not very efficient in filtering the spectrum. Moreover, we consider homogeneous filters $G(\boldsymbol{r})$, independent of $\boldsymbol{x}$ (as an example, the model resolution does not change with the position: note that this constraint is relaxed in recent models with variable resolution).

The velocity components and the scalar fields are decomposed in a filtered part and a residual (often called subgrid-scale) part; for instance the velocity is written as:

$$\tilde{\boldsymbol{u}}(\boldsymbol{x}, t) = \boldsymbol{u}_{\mathrm{F}} + \boldsymbol{u}_{\mathrm{S}} \ , \tag{6.32}$$

where:

$$\boldsymbol{u}_{\mathrm{F}}(\boldsymbol{x}, t) = \int G(\boldsymbol{r})\tilde{\boldsymbol{u}}(\boldsymbol{x}-\boldsymbol{r}, t)\,\mathrm{d}^3\boldsymbol{r} \equiv [\tilde{\boldsymbol{u}}]_{\mathrm{F}} \ , \tag{6.33}$$

with the normalization condition

$$\int G(\boldsymbol{r})\,\mathrm{d}^3\boldsymbol{r} = 1 \ . \tag{6.34}$$

Remember that filtering and differentiation with respect to time and to space commute: see Sect. 2.3.2.3 and Pope (2000, p. 562).

If the velocity is averaged over a box of sides L_{A1}, L_{A2}, L_{A3}, which reproduces the volume averaging (2.41), the corresponding filter is:

$$G(\boldsymbol{r}) = \frac{1}{\Delta}\prod_{i=1}^{3}\mathrm{H}(L_{Ai}/2 - |r_i|)\ , \qquad (6.35)$$

where H is the box function and $\Delta = (L_{A1}L_{A2}L_{A3})^{1/3}$ is the characteristic filter width: see Deardorff (1970) and also Pope (2000, p. 576).

Broadly speaking (broadly because the box filter is non-local in the Fourier space), eddies characterized by wavenumber $k < \Delta^{-1}$ are explicitly represented in the simulation, while those with $k > \Delta^{-1}$ are the subgrid-scale ones.

The decomposition (6.32) is analogous to the Reynolds decomposition, but $\boldsymbol{u}_F$ is a random field (not the first moment of a random field): this is the basic difference between Reynolds averaged fields and filtered fields. Moreover, in general the filtered residual field is different from zero:

$$\int G(\boldsymbol{r})\boldsymbol{u}_S(\boldsymbol{x}-\boldsymbol{r},t)\,\mathrm{d}^3\boldsymbol{r} \neq 0\ , \qquad (6.36)$$

at variance with the fluctuations. However, $[\boldsymbol{u}_S]_F = 0$ if the filter is a projection, i.e.

$$\int G(\boldsymbol{r})\boldsymbol{u}_F(\boldsymbol{x}-\boldsymbol{r},t)\,\mathrm{d}^3\boldsymbol{r} = \boldsymbol{u}_F\ , \qquad (6.37)$$

as can be seen filtering (6.32) (see also Pope 2000, p. 567, Eq. (13.18)).

In many turbulence applications, $l_K < \Delta < L_E$. In meteorological weather forecast models, it is common to chose the (horizontal) sides of the box much larger than L_E, so $\boldsymbol{u}_F \approx \boldsymbol{U}$. With the increasing computation power, simulations of boundary layer (and convection) cases can be done with filter width approaching the integral scale; this leads to changing the parameterization of the subgrid-scale turbulence, see Sect. 6.3.3.

6.3.1 Filtered Equations

The equations for the filtered fields are obtained by applying the filter to NSE.

6.3.1.1 Some Definitions

From the kinetic energy field $[\tilde{u}_i\tilde{u}_i]/2$ the filtered kinetic energy is derived, and splitted into the kinetic energy of the filtered field E_F and the residual (subgrid scale) kinetic energy E_S:

$$\frac{1}{2}[\tilde{u}_i\tilde{u}_i]_\mathrm{F} = E_\mathrm{F} + E_\mathrm{S} , \tag{6.38}$$

where

$$E_\mathrm{F} = \frac{u_{Fi}u_{Fi}}{2} , \tag{6.39}$$

and obviously

$$E_\mathrm{S} = \frac{1}{2}[\tilde{u}_i\tilde{u}_i]_\mathrm{F} - \frac{u_{Fi}u_{Fi}}{2} , \tag{6.40}$$

(see Pope 2000, Eq. (13.121)). Note that the filtered kinetic energy is in general an unsteady inhomogeneous field.

Define the residual stress tensor:

$$\tau_{\mathrm{S}ij} = (u_iu_j)_\mathrm{F} - u_{\mathrm{F}i}u_{\mathrm{F}j} , \tag{6.41}$$

and the modified filtered pressure:

$$\tilde{p}_\mathrm{F} = p_\mathrm{F} + \frac{2}{3}\rho_{00}E_\mathrm{S} . \tag{6.42}$$

The residual kinetic energy can be written:

$$E_\mathrm{S} = \frac{1}{2}\tau_{\mathrm{S}ii} , \tag{6.43}$$

and the (anisotropic) residual stress tensor is defined:

$$\tau_{\mathrm{s}ij} = \tau_{\mathrm{S}ij} - \frac{2}{3}E_\mathrm{S}\delta_{ij} . \tag{6.44}$$

Note that the residual stress tensor is analogous to the Reynolds stress tensor.

Define the filtered rate of strain:

$$S_{\mathrm{F}ij} = \frac{1}{2}\left(\frac{\partial u_{\mathrm{F}i}}{\partial x_j} + \frac{\partial u_{\mathrm{F}j}}{\partial x_i}\right) , \tag{6.45}$$

and note that, with Δ in the inertial subrange, $|\boldsymbol{u}_\mathrm{F}| \sim \varepsilon^{1/3}\Delta^{1/3}$ for a scale length $\sim \Delta$: thus the order of magnitude of $|S_{\mathrm{F}ij}|$ is $\varepsilon^{1/3}\Delta^{-2/3}$ and decreases as the filter width increases (see a detailed evaluation in Pope 2000, Eq. (13.76)).

6.3.1.2 The Continuity Equation

For constant density flows, the continuity equation for the filtered and subgrid velocity fields reads:

$$\frac{\partial u_{\mathrm{F}i}}{\partial x_i} = 0 \,, \tag{6.46}$$

$$\frac{\partial u_{\mathrm{S}i}}{\partial x_i} = 0 \,. \tag{6.47}$$

6.3.1.3 The Filtered Momentum Equation

The filtered momentum equation reads:

$$\frac{\partial u_{\mathrm{F}i}}{\partial t} + u_{\mathrm{F}j}\frac{\partial u_{\mathrm{F}i}}{\partial x_j} = \nu\frac{\partial^2 u_{\mathrm{F}i}}{\partial x_j \partial x_j} - \frac{1}{\rho_{00}}\frac{\partial \tilde{p}_{\mathrm{F}}}{\partial x_i} - \frac{\partial \tau_{\mathrm{s}ij}}{\partial x_j} + \frac{g}{\theta_{00}}\delta_{i3}\Theta_{\mathrm{F}} \,. \tag{6.48}$$

With the characteristic filter width in the inertial subrange but larger than the Kolmogorov scale, we can evaluate the Reynolds number: $R_{\mathrm{e}} \propto \Delta^{4/3}/\nu \gg 1$ if $\Delta \gg \mathrm{l_k}$, so that in (6.48) the viscous diffusive term can be neglected.

6.3.1.4 The Equations for the Kinetic Energy of the Filtered Velocity and for the Residual Kinetic Energy

The equation for E_{F} is obtained by multiplying (6.48) by $u_{\mathrm{F}i}$. It reads:

$$\begin{aligned} \frac{\partial E_{\mathrm{F}}}{\partial t} + u_{\mathrm{F}j}\frac{\partial E_{\mathrm{F}}}{\partial x_j} = \\ \nu\frac{\partial^2 E_{\mathrm{F}}}{\partial x_j \partial x_j} - \frac{u_{\mathrm{F}i}}{\rho_{00}}\frac{\partial \tilde{p}_{\mathrm{F}}}{\partial x_i} - \frac{\partial}{\partial x_j}\left(u_{\mathrm{F}i}\tau_{\mathrm{s}ij}\right) \\ + \frac{g}{\theta_{00}}\delta_{i3}u_{\mathrm{F}i}\Theta_{\mathrm{F}} - \nu\left(\frac{\partial u_{\mathrm{F}i}}{\partial x_j}\right)^2 + \tau_{\mathrm{s}ij}\frac{\partial u_{\mathrm{F}i}}{\partial x_j} \,. \end{aligned} \tag{6.49}$$

In this equation we recognize the viscous diffusion term, the divergence of the third order terms, the forcing due to buoyancy, the dissipation and the term of coupling between the residual stress and the gradient of filtered velocity. As noted discussing (6.48), being Δ larger than the Kolmogorov scale, the Reynolds number is large so the viscous diffusion can be neglected. The dissipation term:

$$\varepsilon_{\mathrm{F}} = \nu\left(\frac{\partial u_{\mathrm{F}i}}{\partial x_j}\right)^2 \,, \tag{6.50}$$

scales on $\nu\varepsilon^{-2/3}\Delta^{-4/3}$ being ε the (true) turbulent dissipation that characterizes the inertial subrange. Thus the ratio $\varepsilon_F/\varepsilon$ can be evaluated as function of the ratio between the filter width and the Kolmogorov scale:

$$\frac{\varepsilon_F}{\varepsilon} \propto \left(\frac{l_K}{\Delta}\right)^{4/3} , \tag{6.51}$$

i.e. the filtered dissipation is (much) smaller than the turbulent dissipation and can be neglected.

The last term can be written as:

$$\tau_{sij}\frac{\partial u_{Fi}}{\partial x_j} \equiv -\mathscr{P}_F = \tau_{sij}S_{Fij} . \tag{6.52}$$

The equation for E_S, deduced by Lilly (1967) and discussed by Wyngaard (2004a, Eq. (L2.9)), reads:

$$\frac{\partial E_S}{\partial t} + u_{Fj}\frac{\partial E_S}{\partial x_j} = \frac{\partial \mathscr{T}}{\partial x_j} - \varepsilon_S + \mathscr{P}_F , \tag{6.53}$$

where $\mathscr{T}$ indicates the third-order terms, the dissipation $\varepsilon_S = [\nu(\partial u_i/\partial x_j)^2]_F - \varepsilon_F \approx \varepsilon$ is well approximated by the dissipation averaged over the box, which in homogeneous conditions is the same of the turbulent dissipation (2.74). Here the term related to the heat flux is neglected, assuming that it acts mainly on the filtered field (on the larger scales).

6.3.2 *Closure of the Filtered Equations*

Different approaches are obviously possible to close the equations: here we will present two among the others, widely used in atmospheric applications.

6.3.2.1 The Smagorinsky (1963) Model

The simplest model for the anisotropic stress tensor was proposed by Smagorinsky (1963):

$$\tau_{sij} = -2K_\Delta S_{Fij} \tag{6.54}$$

with the model for the eddy diffusion coefficient:

$$K_\Delta = l_\Delta^2 \left(S_{Fij}S_{Fij}\right)^{1/2} = (C_S\Delta)^2 \left(S_{Fij}S_{Fij}\right)^{1/2} \tag{6.55}$$

where l_Δ is a length scale which is assumed to be proportional to the filter width. With Δ in the inertial subrange, from the scaling $l_\Delta \sim \Delta$ and $S_{Fij} \sim \varepsilon^{1/3}\Delta^{-2/3}$ the eddy diffusion coefficient scales as $K_\Delta \sim \varepsilon^{1/3}\Delta^{4/3}$ (remember the diffusivity for absolute dispersion processes). Moreover, C_S must be constant (independent of Δ): it can be evaluated equating $\mathscr{P}_F$ and ε, that balance approximately in steady conditions, see (6.53). A detailed computation can be found in Pope (2000, Eq. (13.132)), giving $C_S \approx 0.17$ for high Reynolds number free flows. Increasing Δ the eddy diffusion coefficient K_Δ increases and saturates to the value K_τ given by (6.3), which is expected to be different for vertical and horizontal directions.

The rate of transfer of energy from filtered to residual motion is

$$\mathscr{P}_F = 2K_\Delta S_{Fij}S_{Fij} , \tag{6.56}$$

so for any eddy viscosity model with positive K_Δ the transfer of energy is always from filtered (resolved) to residual motion.

6.3.2.2 The Germano Dynamic Model for the Subgrid Scale Motion

A problem with the (6.54) and (6.55) model is that the value of C_S changes in different flows: in wall flows for instance is smaller than in free flow conditions.

Thus the model has been modified by Germano et al. (1991a,b) using the eddy diffusion concept dynamically applied, thus allowing for a variability of the constant, which is computed locally. Further improvements of the model can be found in the literature.

The fundamental idea is to use a test filter, with resolution coarser than the resolution of the simulation but still in the inertial subrange, to apply to the unresolved stresses (for both the filtered fields) the eddy diffusion closure and to evaluate C_S from the knowledge of the subgrid scale stresses of the test filtered fields, given by the higher resolution simulation. A detailed description can be found in Pope (2000, Sect. 13.6.1).

6.3.3 The Transition from RANS to LES

Increasing Δ, the filtered velocity tends, on average, to the mean velocity and the residual velocity tends to the fluctuation: $\boldsymbol{u}_F = \boldsymbol{U}$ and $\boldsymbol{u}_S = \boldsymbol{u}$; the residual stress tensor tends to the Reynolds stress tensor. We highlight that this convergence occurs on average (averaging $\boldsymbol{u}_F$ over many realizations): as noted by Wyngaard (2004b), for $\Delta/L_E = 1$ the $\boldsymbol{u}_F$ field is a random field, while the ensemble averaged $\boldsymbol{U}$ is not, in spite of the fact that are relative to the same box.

Let consider a simple homogeneous shear flow. The residual shear stress reads:

$$\tau_{s12} = -K_{\Delta}\frac{du_{F1}}{dx_2} = -l_{\Delta}^2\left|\frac{du_{F1}}{dx_2}\right|\frac{du_{F1}}{dx_2} \tag{6.57}$$

and the Reynolds stress reads:

$$\langle u_1 u_2\rangle = -K_{\tau}\frac{dU_1}{dx_2} = -l_{\tau}^2\left|\frac{dU_1}{dx_2}\right|\frac{dU_1}{dx_2} \tag{6.58}$$

Evidently, for $\Delta/L_E \geq 1$ the eddy diffusion coefficients are equal: $K_{\Delta} = K_{\tau}$ and the Smagorinsky length equals the mixing length $l_{\Delta} = l_{\tau}$; thus $C_S\Delta = l_{\tau}$. Because l_{τ} is independent of Δ, it results that C_S is dependent on the resolution.

This discussion highlights the fact that the parameterization of turbulence changes from models with a filter size in the inertial sublayer to models with filter size larger than the integral scale.

6.4 Numerical Simulations of PBL Problems

There is a overwhelming number of papers dealing with the numerical simulations of boundary layer problems and presenting comparisons among models and data. New results appear continuously. Here we shall discuss a few exemplar cases.

- The diurnal evolution of the PBL. Extensive model comparisons are reported by Cuxart et al. (2006) (dealing with RANS models) and by Beare et al. (2006) (dealing with LES), both referring to the same case of SBL; simulations of the complete diurnal cycle are discussed by Svensson et al. (2011), using CASES99 data set, and by Bosveld et al. (2014a,b), using CABAUW data set. See also the discussion by Holtslag et al. (2013). Some conclusions can be drawn:

  the higher-order closures (based on the TKE equation) perform better as far as the simulation of the second-order moments (the fluxes) is concerned, but the first order moments can be equally well simulated by simpler (less expensive) closures. A big importance lies in the tuning of the model constants, which can be rephrased saying that we are far from a true description of the physics of turbulence;

  the SBL depth is a fuzzy concept; the different estimates reflect the model differences and also its definition; it is a concept that must be handled carefully;

  LES are not necessarily more accurate than RANS in the description of all the features of (atmospheric) turbulence;
- The flow over topography. Concerning the flow over isolated topography, the lesson from the RANS simulations, whose prototype is the work by Zeman and Jensen (1987), is the need to account in the parameterizations for modifications of turbulence due to the deformation of the eddies. Further comparisons (Bechmann

and Sorensen 2010; Bechmann et al. 2011) have shown that LES are suited to describe the wake (characterized by large scale eddies) and less suited to simulate the speed-up at the hilltop, in the inner layer, because the resolution should be very high near the ground. Flows over complex topography, including breezes and other typical phenomena characteristic of the mountain micrometeorology, are subject of large applicative interest and need accurate description of boundary layer turbulence. A discussion can be found in Rotach and Zardi (2007): the heterogeneity of the terrain (its shape, the different radiation on the slopes) needs special attention. Relations that hold locally cannot be extended globally: for instance, the PBL depth turns out to be strictly related to the height of the surrounding mountains. Novel applications of numerical weather prediction models at high resolution in order to solve the features of the boundary layer are increasing: see as an example Seaman et al. (2012).
- The flow within the urban canopy. The street canyons have been investigated since many years ago; in general, the flow around buildings has an increasing importance for environmental and building design purposes. To this end, LES are the most useful tool to depict the large scale eddies that develop among the buildings, the turbulent wakes, and the thermal effects related to differential heating of the surfaces. Recent papers (see, for instance Park et al. 2015) contain references to previous work and show the direction of the research towards even more detailed description of the geometry of the city.
- Dispersion simulations. Concerning the absolute dispersion problem, the LSM described in Sects. 5.6.2 and 5.6.3, and their Eulerian counterpart (essentially, the diffusion equation) require as input the velocity field probability density function, or some statistical moments. RANS and LES can give the necessary input: while RANS can be used strictly to represent the ensemble average of the flow field, LES provide single realizations of the flow. As noted in Sect. 6.3.3, because even at the same resolution the two approaches give velocity fields with different characteristics, their use to drive LSM will describe different aspects of the dispersion, along the line of the meandering discussion (Sect. 5.4).

Besides the applications discussed up to now, it is worth to remember that numerical simulations are used to produce data sets suited to investigate specific problems, possibly in parallel to the experimental research, as stated in Chap. 1. Two cases are cited here.

- PBL research via LES. To integrate available observations and to explore conditions difficult to observe, LES are suitable tools for producing extensive data bases of geophysical relevance. It is worth mentioning the DATABASE64 (Esau 2009) which contains numerical simulations of PBL and has been used to test new models (for instance Zilitinkevich et al. 2013).
- Properties of pair dispersion. DNS and LES are widely used to compute turbulent velocity fields in order to study dispersion of pair of tracer parcels or of inertial particles (see Toschi and Bodenschatz 2009). We have already cited the use of DNS for investigating inertial particle dispersion (Sect. 5.7); thanks to the dominant role of the large eddies, LES are used to study the relative

dispersion properties in strongly non-isotropic turbulence conditions (Mazzitelli and Lanotte 2012; Mazzitelli et al. 2014).

Exercises

6.1 Derive the algebraic equations for the second-order moments in the horizontal, homogeneous, steady state approximation.

References

R.J. Beare, M.K. Macvean, A.A.M. Holtslag, J. Cuxart, I.N. Esau, J.-C. Golaz, M.A. Jimenez, M. Khairoutdinov, B. Kosovic, D. Lewellen, T.S. Lund, J.K. Lundquist, A. Mccabe, A.F. Moene, Y. Noh, S. Raasch, P.P. Sullivan, An intercomparison of Large-Eddy Simulations of the stable boundary layer. Bound.-Layer Meteorol. **118**, 247–272 (2006)

A. Bechmann, N.N. Sorensen, Hybrid RANS/LES method for wind flow over complex terrain. Wind Energy **13**, 36–50 (2010)

A. Bechmann, N.N. Sorensen, J. Berg, J. Mann, P.E. Réthoré, The Bolund experiment, part II: blind comparison of microscale flow models. Bound.-Layer Meteorol. **141**, 245–271 (2011)

A.K. Blackadar, The vertical distribution of wind and turbulence exchange in a neutral atmosphere. J. Geophys. Res. **62**, 3095–3102 (1962)

F.C. Bosveld, P. Baas, G.J. Steeneveld, A.A.M. Holtslag, W.M. Angevine, E. Bazile, H. De Bruin, D. Deacu, J. Edwards, M. Ek, V.E. Larson, M. Raschendorfer, G. Svensson, The third GABLS intercomparison case for evaluation studies of boundary-layer models. Part B: results and process understanding. Bound.-Layer Meteorol. **152**, 157–187 (2014a)

F.C. Bosveld, P. Baas, E. van Meijgaard, E.I.F. de Bruijn, G.J. Steeneveld, A.A.M. Holtslag, The Third GABLS intercomparison case for evaluation studies of boundary-layer models. Part A: case selection and set-up. Bound.-Layer Meteorol. **152**, 133–156 (2014b)

J. Boussinesq, Essai sur la theorie des eaux courantes. Mem. Pres. par div. savants a l'Acad. Sci. (Paris) **23**, 46 (1877)

A.R. Brown, L.M. Grant, Non-local mixing of momentum in the convective boundary layer. Bound.-Layer Meteorol. **84**, 1–22 (1997)

A.R. Brown, R.J. Beare, J. Edwards, A. Lock, S.J. Keogh, S.F. Milton, D.N. Walters, Upgrades to the boundary-layer scheme in the met office numerical weather prediction model. Bound.-Layer Meteorol. **128**, 117–132 (2008)

A. Buzzi, M. Fantini, P. Malguzzi, P. Nerozzi, Validation of a limited area model in cases of Mediterranean cyclogenesis: surface fields and precipitation scores. Meteorol. Atmos. Phys. **53**, 137–153 (1994)

A. Buzzi, M. D'Isidoro, S. Davolio, A case-study of an orographic cyclone South of the Alps during the MAP SOP. Q. J. R. Meteorol. Soc. **129**, 1795–1818 (2003)

V. Canuto, Turbulent convection with overshooting-Reynolds stress approach. Astrophys. J. **392**, 218–232 (1992)

J. Cuxart, A.A.M. Holtslag, R.J. Beare, E. Bazile, A. Beljaars, A. Cheng, L. Conangla, M. Ek, F. Freedman: Single-column model intercomparison for a stably stratified atmospheric boundary layer. Bound.-Layer Meteorol. **118**, 273–303 (2006)

J.W. Deardorff, The counter-gradient heat flux in the atmosphere and in the laboratory. J. Atmos. Sci. **23**, 503–506 (1966)

J.W. Deardorff, A numerical study of three-dimensional turbulent channel flow at large reynolds numbers. J. Fluid Mech. **41**, 453–465 (1970)
I. Esau, Large-eddy simulations of geophysical turbulent flows with applications to planetary boundary layer research (2009) arXiv preprint arXiv:0907.0103
E. Ferrero, L. Quan, D. Massone, Turbulence in the stable boundary layer at higher richardson numbers. Bound.-Layer Meteorol. **139**, 225–240 (2011)
M. Germano, U. Piomelli, P. Moin, W.H. Cabot, A dynamic subgrid-Řscale eddy viscosity model. Phys. Fluids A **3**, 1760–1765 (1991a)
M. Germano, U. Piomelli, P. Moin, W.H. Cabot, Erratum: a dynamic subgrid-scale eddy viscosity model. Phys. Fluids **A 3**, 3128 (1991b)
V. Gryanik, J. Hartmann, A turbulence closure for the convective boundary layer based on a two-scale mass-flux approach. J. Atmos. Sci. **59**, 2729–2744 (2002)
A.A.M. Holtslag, C.-H. Moeng, Eddy diffusivity and countergradient transport in the convective atmospheric boundary layer. J. Atmos. Sci. **48**, 1690–1698 (1991)
A.A.M. Holtslag, G. Svensson, P. Baas, R.J. Beare, A. Beljaars, F.C. Bosveld, J. Cuxart, J. Iindvall, G.J. Steeneveld, M. Tjernström, B.J.H. van de Wiel, Stable atmospheric boundary layers and diurnal cycles. Bull. Am. Meteorol. Soc. **94**, 1691–1705 (2013)
A.N. Kolmogorov, The equations of turbulent motion in an incompressible fluid. Izvestia Acad. Sci. USSR; Phys. **6**, 56–58 (1942)
D.K. Lilly, The representation of small scale turbulence in numerical simulation experiments, in *IBM Scientific Computing Symposium on environmental sciences*, Yorktown Heights, 1967
T. Mauritsen, G. Svensson, S.S. Zilitinkevich, I.N. Esau, L. Enger, B. Grisogono, A total turbulent energy closure model for neutrally and stably stratified atmospheric boundary layers. J. Atmos. Sci. **64**, 4113–4126 (2007)
I. Mazzitelli, A.S. Lanotte, Active and passive scalar intermittent statistics in turbulent atmospheric convection. Physica D: Nonlinear Phenom. **241**, 251–259 (2012)
I. Mazzitelli, F. Fornarelli, A. Lanotte, P. Oresta, Pair and multi-particle dispersion in numerical simulations of convective boundary layer turbulence. Phys. Fluids **26**, 055110 (2014)
G.L. Mellor, T. Yamada, A hierarchy of turbulence closure models for planetary boundary layers. J. Atmos. Sci. **31**, 1791–1806 (1974)
G.L. Mellor, T. Yamada, Development of a turbulence closure model for geophysical fluid problems. Rev. Geophys. Space Phys. **20**, 851–875 (1982)
M.D. Millionshchikov, Theory of homogeneous isotropic turbulence. Dokl. Akad. Nauk SSSR **32**, 611–614 (1941)
D.V. Mironov, Turbulence in the lower troposphere: second-order closure and mass-flux modelling frameworks, in *Interdisciplinary Aspects of Turbulence*, vol. 756, pp. 1–61(Springer, Berlin, Heidelberg, 2009), ed. by W. Hillebrandt, F. Kupka
A.S. Monin, A.M. Yaglom, *Statistical Fluid Mechanics*, vol. I, 769 pp. (MIT Press, Cambridge, 1971)
A.S. Monin, A.M. Yaglom, *Statistical Fluid Mechanics*, vol. II, 874 pp. (MIT Press, Cambridge, 1975)
S.-B. Park, J. J. Baik, B.-S. Han, Large-eddy simulation of turbulent flow in a densely built-up urban area. Environ. Fluid Mech. **15**, 235–250 (2015)
S. Pope, *Turbulent Flows* (Cambridge University Press, Cambridge, 2000)
L. Prandtl, Über ein neues Formelsystem für die ausgebildete Turbulenz. Nachr. Akad. Wiss. Goettingen Math-Phys. **K1**, 6–19 (1945)
M.W. Rotach, D. Zardi, On the boundary-layer structure over highly complex terrain: key findings from map. Quart. J. R. Meteorol. Soc. **133**, 937–948 (2007)
J.C. Rotta, Statisthe Theorie nichthomogener Turbulenz. Z. Phys. **129**, 547–572 (1951)
P. Sagaut, *Large Eddy Simulation for Incompressible Flows* (Springer, Berlin, Heidelberg, 1998)
B.L. Sawford, F.M. Guest, Lagrangian stochastic analysis of flux-gradient relationships in the convective boundary layer. J. Atmos. Sci. **44**, 1152–1165 (1987)

N.L. Seaman, B.J. Gaudet, D.R. Stauffer, L. Mahrt, S.J. Richardson, J.R. Zielonka, J.C. Wyngaard, Numerical prediction of submesoscale flow in the nocturnal stable boundary layer over complex terrain. Mon. Weater Rev. **140**, 956–977 (2012)

J. Smagorinsky, General circulation experiments with the primitive equations: I. The basic experiment. Mon. Weather Rev. **91**, 99–164 (1963)

G. Svensson, A.A.M. Holtslag, V. Kumar, T. Mauritsen, G.J. Steeneveld, W.M. Angevine, E. Bazile, A. Beljaars, H. De Bruin, A. Cheng, L. Conangla, J. Cuxart, M. Ek, M.J. Falk, F. Freedman, H. Kitagawa, V.E. Larson, A. Lock, J. Mailhot, V. Masson, S. Park, J. Pleim, S. Soederberg, W. Weng, M. Zampieri, Evaluation of the diurnal cycle in the atmospheric boundary layer over land as represented by a variety of single-column models:the second GABLS experiment. Bound.-Layer Meteorol. **140**, 177–206 (2011)

F. Toschi, E. Bodenschatz, Lagrangian properties of particles in turbulence. Annu. Rev. Fluid Mech. **41**, 375–404 (2009)

J.C. Wyngaard, A physical mechanism for the asymmetry in top-down and bottom-up diffusion. J. Atmos. Sci. **44**, 1083–1087 (1987)

J.C. Wyngaard, Changing the face of small-scale meteorology, in *Atmospheric Turbulence and Mesoscale Meteorology*, pp. 17–34 (Cambridge University Press, Cambridge, 2004a) ed. by E. Fedorovich, R. Rotunno, B. Stevens

J.C. Wyngaard, Toward numerical modelling in the 'terra incognita'. J. Atmos. Sci. **61**, 1816–1826 (2004b)

J.C. Wyngaard, *Turbulence in the Atmosphere* (Cambridge University Press, Cambridge, 2010)

M. Zampieri, P. Malguzzi, A. Buzzi, Sensitivity of quantitative precipitation forecasts to boundary layer parameterization: a flash flood case study in the Western Mediterranean. Nat. Hazards Earth Syst. Sci. **5**, 603–612 (2005)

O. Zeman, N.O. Jensen, Modification of turbulence characteristics in flow over hills. Quart. J. R. Meteor. Soc. **113**, 55–80 (1987)

S.S. Zilitinkevich, T. Elperin, N. Kleeorin, I. Rogachevskii, I.N. Esau, A hierarchy of Energy- and Flux-Budget (EFB) turbulence closure models for stably-stratified geophysical flows. Bound.-Layer Meteorol. **146**, 341–373 (2013)

S.S. Zilitinkevich, V.M. Gryanik, V.N. Lykossov, D.V. Mironov, Third-order transport and non-local turbulence closures for convective boundary layers. J. Atmos. Sci. **56**, 3463–3477 (1999)

Solutions

Exercises of Chap. 2

2.1 Let consider a small departure $\tilde{p}$ from the hydrostatic pressure and the corresponding departures of the density $\tilde{\rho}$ and of the temperature $\tilde{T}$. Thus, using the state equation Eq. (2.5):

$$\tilde{\rho} = \frac{\partial \rho_{\mathrm{f}}}{\partial T_f}|_{\mathrm{a}} \tilde{T} + \frac{\partial \rho_{\mathrm{f}}}{\partial p_f}|_{\mathrm{a}} \tilde{p} = -\frac{\rho_{\mathrm{a}}}{T_{\mathrm{a}}} \tilde{T} + \frac{\tilde{p}}{R\,T_{\mathrm{a}}} \,. \tag{S2.1}$$

When buoyancy and diffusivity are negligible, pressure $\tilde{p}$ is of order of magnitude of $\rho_{00}\tilde{u}^2$. Substituting in Eq. (S2.1) it results:

$$\tilde{\rho} = -\frac{\rho_{\mathrm{a}}}{T_{\mathrm{a}}} \tilde{T} + \frac{\rho_{00}\tilde{u}^2}{R\,T_{\mathrm{a}}} \,, \tag{S2.2}$$

and

$$\frac{\tilde{\rho}}{\rho_{\mathrm{a}}} \approx -\frac{\tilde{T}}{T_{\mathrm{a}}} + \frac{\tilde{u}^2}{R\,T_{\mathrm{a}}} \,. \tag{S2.3}$$

Using $T_{\mathrm{a}} = 273\,\mathrm{K}$, $\tilde{T} = 1\,\mathrm{K}$ and $\tilde{u} = 10\,\mathrm{m\,s^{-1}}$ as estimates of the temperature and velocity near the ground, it results that the second term at RHS is smaller than the first one (this term is important as the velocity is of the order of the sound speed). Thus:

$$\frac{\tilde{\rho}}{\rho_{\mathrm{a}}} \simeq -\frac{\tilde{T}}{T_{\mathrm{a}}} \simeq -\frac{\tilde{\theta}}{\theta_{\mathrm{a}}} \,. \tag{S2.4}$$

F. Tampieri, *Turbulence and Dispersion in the Planetary Boundary Layer*,
Physics of Earth and Space Environments, DOI 10.1007/978-3-319-43604-3

2.2 Writing the material derivative for $\tilde{q}_\mathrm{c} = \tilde{c}/\rho_\mathrm{f}$ and computing the partial derivatives:

$$\frac{\mathrm{D}\tilde{q}_\mathrm{c}}{\mathrm{D}t} \equiv \frac{\partial \tilde{q}_\mathrm{c}}{\partial t} + \tilde{u}_i \frac{\partial \tilde{q}_\mathrm{c}}{\partial x_i} = \frac{1}{\rho_\mathrm{f}} \frac{\partial \tilde{c}}{\partial t} - \frac{\tilde{c}}{\rho_\mathrm{f}^2} \frac{\partial \rho_\mathrm{f}}{\partial t} + \frac{\tilde{u}_i}{\rho_\mathrm{f}} \frac{\partial \tilde{c}}{\partial x_i} - \frac{\tilde{c}\,\tilde{u}_i}{\rho_\mathrm{f}^2} \frac{\partial \rho_\mathrm{f}}{\partial x_i} . \tag{S2.5}$$

Using (2.14) divided by ρ_f and (2.10) multiplied by $\tilde{c}/\rho_\mathrm{f}^2$ it results:

$$\frac{\mathrm{D}\tilde{q}_\mathrm{c}}{\mathrm{D}t} = \frac{\kappa_\mathrm{c}}{\rho_\mathrm{f}} \frac{\partial^2 \tilde{c}}{\partial x_i\, \partial x_i} , \tag{S2.6}$$

i.e. the mixing ratio is conserved if molecular mixing is negligible.

2.3 From (2.21), dividing by c_p and integrating from z_r to z, it results:

$$\ln \frac{\theta_\mathrm{f}(z)}{\theta_\mathrm{f}(z_\mathrm{r})} = \ln \frac{T_\mathrm{f}(z)}{T_\mathrm{f}(z_\mathrm{r})} + \ln \left(\frac{p_\mathrm{f}(z)}{p_\mathrm{f}(z_\mathrm{r})} \right)^{-R/c_\mathrm{p}} , \tag{S2.7}$$

from which (2.22) results. To show that the potential temperature is constant if the temperature gradient is adiabatic, the vertical derivative of (2.22) is computed:

$$\frac{\mathrm{d}\theta_\mathrm{f}}{\mathrm{d}z} = \frac{\mathrm{d}T_\mathrm{f}}{\mathrm{d}z} \left(\frac{p_\mathrm{f}}{p_\mathrm{f}(z_\mathrm{r})} \right)^{-R/c_\mathrm{p}} - T_\mathrm{f} \frac{R}{c_\mathrm{p}} \left(\frac{p_\mathrm{f}}{p_\mathrm{f}(z_\mathrm{r})} \right)^{-R/c_\mathrm{p}} \frac{1}{p_\mathrm{f}} \frac{\mathrm{d}p_\mathrm{f}}{\mathrm{d}z} , \tag{S2.8}$$

and using (2.6) and (2.19) written for a vertical motion, i.e. substituting the material derivative with the derivative with respect to z, it results that $\mathrm{d}\theta_\mathrm{f}/\,\mathrm{d}z = 0$.

2.4 Applying Leibniz theorem:

$$\frac{\partial \langle \tilde{a} \rangle}{\partial \vartheta} = \frac{1}{T_\mathrm{A}} \left[\langle \tilde{a} \rangle(\vartheta + T_\mathrm{A}/2) - \langle \tilde{a} \rangle(\vartheta - T_\mathrm{A}/2) \right] + \frac{1}{T_\mathrm{A}} \int_{\vartheta - T_\mathrm{A}/2}^{\vartheta + T_\mathrm{A}/2} \frac{\partial \langle \tilde{a} \rangle(t)}{\partial \vartheta} \,\mathrm{d}t , \tag{S2.9}$$

and the last integral is zero. Also:

$$\left\langle \frac{\partial \tilde{a}}{\partial t} \right\rangle \equiv \frac{1}{T_\mathrm{A}} \int_{\vartheta - T_\mathrm{A}/2}^{\vartheta + T_\mathrm{A}/2} \frac{\partial \tilde{a}}{\partial t} \,\mathrm{d}t = \frac{1}{T_\mathrm{A}} \left[\tilde{a}(\vartheta + T_\mathrm{A}/2) - \tilde{a}(\vartheta - T_\mathrm{A}/2) \right] . \tag{S2.10}$$

Thus the commutative property holds.

2.5 Consider a stationary process and take the average of a generic function f. The filtered function is:

$$f^{(F)}(x) = \int_{-\infty}^{\infty} G(r) f(x - r) \,\mathrm{d}r , \tag{S2.11}$$

where G is given by (2.59). The covariance for the filtered function, from the definition (2.42) at fixed position (here the independent variables are assumed to be times), reads:

$$R^{(F)}(|x-z|) = \left\langle f^{(F)}(x) f^{(F)}(z) \right\rangle \int_{-\infty}^{\infty}\int_{-\infty}^{\infty} G(r)G(t)\langle f(x-r)f(z-t)\rangle \, \mathrm{d}r\, \mathrm{d}t \,, \tag{S2.12}$$

thus

$$R^{(F)}(|x-z|) = \int_{-\infty}^{\infty}\int_{-\infty}^{\infty} G(r)G(t)R(x-z-r+t)\, \mathrm{d}r\, \mathrm{d}t \,. \tag{S2.13}$$

From Eq. (2.55) written for variables k and r

$$E^{(F)}(k) = \frac{1}{\pi}\int_{-\infty}^{\infty} R^{(F)}(a)\mathrm{e}^{-\imath ka}\, \mathrm{d}a =$$

$$= \frac{1}{\pi}\int_{-\infty}^{\infty} G(r)\int_{-\infty}^{\infty} G(t)\int_{-\infty}^{\infty} R(a-r+t)\mathrm{e}^{-ika}\, \mathrm{d}a\, \mathrm{d}t\, \mathrm{d}r \tag{S2.14}$$

where $a = |x-z|$. For fixed r and t define $b = a - r + t$; (S2.13) reads:

$$E^{(F)}(k) = \frac{1}{\pi}\int_{-\infty}^{\infty} G(r)\mathrm{e}^{-\imath kr}\int_{-\infty}^{\infty} G(t)\mathrm{e}^{\imath kt}\int_{-\infty}^{\infty} R(b)\mathrm{e}^{-\imath kb}\, \mathrm{d}b\, \mathrm{d}t\, \mathrm{d}r =$$

$$= E(k)\int_{-\infty}^{\infty} G(r)\mathrm{e}^{-\imath kr}\, \mathrm{d}r \int_{-\infty}^{\infty} G(t)\mathrm{e}^{\imath kt}\, \mathrm{d}t \tag{S2.15}$$

Because G is an even function, its Fourier transform is real:

$$\hat{G} = \frac{1}{2\pi}\frac{\sin(\omega T_{\mathrm{A}}/2)}{(\omega T_{\mathrm{A}}/2)} \,, \tag{S2.16}$$

and $\hat{G} = \hat{G}^{*}$; thus finally

$$E^{(F)}(k) = (2\pi)^2 \left|\hat{G}(k)\right|^2 E(k) \tag{S2.17}$$

2.6 We want to compute the integral:

$$I = \int_0^t \mathrm{d}t_1 \int_0^t \mathrm{d}t_2 a(t_1, t_2) \,, \tag{S2.18}$$

where $a(t_1, t_2) = a(|t_1 - t_2|)$.

Define $s = t_2 - t_1$ and $r = (t_2 + t_1)/2$: see Fig. S2.1 for a graphical scheme of the integration domain.

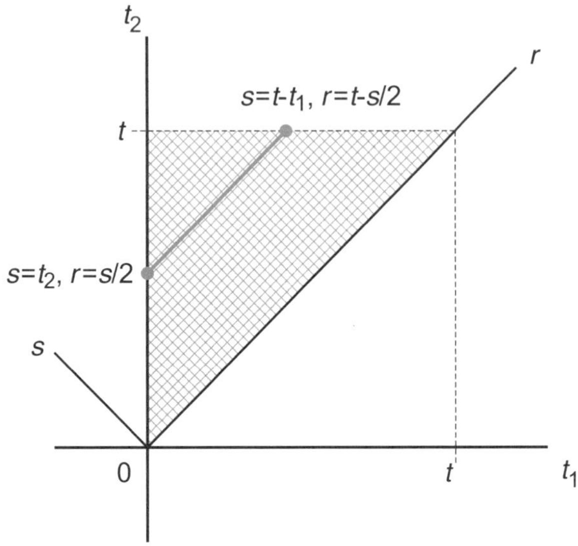

Fig. S2.1 Scheme for the integral (S2.18): for every value of $s = t_2 \equiv t - t_1$ the integral in r is made from $s/2$ to $t - s/2$. Along the *gray line*, the function a is constant

Making at first the integral over r:

$$\int_0^t \mathrm{d}t_1 \int_0^t \mathrm{d}t_2 a(t_2 - t_1) =$$

$$\int_0^t a(s)\,\mathrm{d}s \int_{s/2}^{t-s/2} \mathrm{d}r + \int_{-t}^0 a(s)\,\mathrm{d}s \int_{-s/2}^{t+s/2} \mathrm{d}r =$$

$$\int_0^t a(s)(t-s)\,\mathrm{d}s + \int_{-t}^0 a(s)(t+s)\,\mathrm{d}s =$$

$$\int_0^t a(s)(t-s)\,\mathrm{d}s + \int_0^t a(-s)(t-s)\,\mathrm{d}s\ . \tag{S2.19}$$

Thus:

$$I = 2\int_0^t a(s)(t-s)\,\mathrm{d}s \tag{S2.20}$$

2.7 Given the relation:

$$D(r) = 2\int_0^\infty [1 - \cos(kr)]\,E(k)\,\mathrm{d}k\ , \tag{S2.21}$$

if

$$D(r) = Ar^\gamma, A > 0, 0 < \gamma < 2\ , \tag{S2.22}$$

it results:

$$E(k) = \frac{C}{k^{\gamma+1}} , \tag{S2.23}$$

with:

$$C = \frac{A}{2\int_0^\infty \frac{1-\cos x}{x^{\gamma+1}}\,\mathrm{d}x} = \frac{\Gamma(\gamma+1)\sin(\pi\gamma/2)}{\pi}A . \tag{S2.24}$$

(See, for instance, Monin and Yaglom 1975, p. 90)

2.8 From Olesen et al. (1984, Eq. (1a)):

$$E_{\mathrm{uu}}(k) = \frac{a}{(1+bk)^{5/3}} , \tag{S2.25}$$

whose integral is:

$$\int_{k_1}^{k_2} E_{\mathrm{uu}}(k)\,\mathrm{d}k = \frac{3a}{2b}\left[(1+bk_1)^{-2/3} - (1+bk_2)^{-2/3}\right] . \tag{S2.26}$$

Neglecting viscosity, i.e. $\nu = 0$ and thus $k_2 = \infty$, the coefficients a and b are determined from (2.57) and assuming consistency with (2.103) for large wavenumbers. It results:

$$\int_0^\infty E_{\mathrm{uu}}(k)\,\mathrm{d}k \equiv \frac{3a}{2b} = \langle u^2\rangle , \tag{S2.27}$$

and

$$\frac{a}{b^{5/3}} = C_\varepsilon \varepsilon^{2/3} , \tag{S2.28}$$

where $C_\varepsilon = 0.249 C_{\mathrm{K}}$.

The spectrum reads:

$$E_{\mathrm{uu}}(k) = \frac{C_\varepsilon \varepsilon^{2/3}}{(b^{-1}+k)^{5/3}} , \tag{S2.29}$$

where b is a length scale:

$$b = C_{\mathrm{b}}\frac{\langle u^2\rangle^{3/2}}{\varepsilon} , \tag{S2.30}$$

and $C_{\mathrm{b}} = [2/(3C_\varepsilon)]^{3/2} \approx 1.55$.

Using (2.55) for $\chi = 0$ it results the following relation between the integral length scale L_{uu} from the covariance function and the parameter b:

$$b = \frac{3L_{uu}}{\pi} . \tag{S2.31}$$

From (S2.31) and (S2.30) a relation consistent with (2.101) results: the difference in the numerical coefficient arises from the different forms of the covariance function implied by the different spectra.

Exercises of Chap. 3

3.1 From (3.11) we can separate the contribution to the variation of the turbulent potential energy due to the mean temperature profile and to the fluctuations:

$$\delta E_{P} \equiv \langle \delta E_{P} \rangle + \delta E'_{P} = \frac{g}{\theta_{00}} \int_0^l (\Theta + \theta)\, dz . \tag{S3.1}$$

Using the estimate of the temperature fluctuation

$$\theta = \frac{d\Theta_f}{dz} z , \tag{S3.2}$$

which derives from an argument analogous to that discussed in Sect. 2.4.6, the term relative to the fluctuations in (S3.1) can be integrated:

$$\delta E'_{P} = \frac{g}{\theta_{00}} \frac{d\Theta_f}{dz} \frac{l^2}{2} . \tag{S3.3}$$

Taking the average of (S3.3) and substituting l^2 with $\theta^2/(d\Theta_f/dz)^2$ from (S3.2), the definition of TPE results:

$$E_{P} \equiv \langle \delta E'_{P} \rangle = \frac{1}{2} \left(\frac{g}{\theta_{00} N} \right)^2 \langle \theta^2 \rangle \tag{S3.4}$$

3.2 The mean wind profile is obtained by integration of (3.39)–(3.41), while the mean temperature profile from (3.48) and (3.49). Define:

$$F_i = \int_{z_1}^{z_2} \frac{\Phi}{z} dz , \tag{S3.5}$$

where Φ can be either Φ_U or Φ_Θ, and the values of the coefficients α_i are chosen accordingly.

It results:

$$F_0(z_1, z_2, \alpha_0) = \alpha_0 \ln \frac{z_2}{z_1}, \; z_i < \zeta_A \;, \tag{S3.6}$$

$$F_1(\zeta_1, \zeta_2, \alpha_1) = 3\alpha_1 \left(\zeta_1^{-1/3} - \zeta_2^{-1/3} \right), \; \zeta_A < z_i < \zeta_B \;, \tag{S3.7}$$

$$F_2(\zeta_1, \zeta_2, \alpha_2) = 3\alpha_2 \left(\zeta_2^{1/3} - \zeta_1^{1/3} \right), \; z_i > \zeta_B \;, \tag{S3.8}$$

where $\zeta = -z/L$ and the ranges of application of the integrals are given by the interception of the power laws:

$$\zeta_A = \left(\frac{\alpha_1}{\alpha_0} \right)^3 \;, \tag{S3.9}$$

and

$$\zeta_B = \left(\frac{\alpha_1}{\alpha_2} \right)^{3/2} \;. \tag{S3.10}$$

So, for instance, if $z_1 < \zeta_A L$ and $z_2 > \zeta_B L$, the wind profile reads:

$$\kappa \frac{U(z_2) - U(z_1)}{u_*} = F_0(\zeta_1, \zeta_A, 1) + F_1(\zeta_A, \zeta_B, \alpha_1) + F_2(\zeta_B, \zeta_2, \alpha_2) \;. \tag{S3.11}$$

Similarly, if $z_1 < \zeta_A L$ and $z_2 > \zeta_A L$, the temperature profile reads:

$$\kappa \frac{\Theta(z_2) - \Theta(z_1)}{-\theta_*} = F_0(\zeta_1, \zeta_A, \alpha_0) + F_1(\zeta_A, \zeta_2, \alpha_1) \;. \tag{S3.12}$$

3.3 The integral of (3.42) can be done by putting $x = (1 + \alpha_1 \zeta)^{1/3}$. It results (Gradshteyn and Ryzhik 1980):

$$\begin{aligned} \int_{z_1}^{z_2} \Phi_U / z \, dz &= -3 \int_{x_1}^{x_2} \frac{x}{1 - x^3} \, dx \\ = \frac{1}{2} &\left[\ln \frac{(1 - x_2)^2}{1 + x_2 + x_2^2} - \ln \frac{(1 - x_1)^2}{1 + x_1 + x_1^2} \right] \\ + \sqrt{3} &\left[\arctan \frac{2x_2 - 1}{\sqrt{3}} - \arctan \frac{2x_1 - 1}{\sqrt{3}} \right] \;. \end{aligned} \tag{S3.13}$$

3.4 The problem is solved using complex variables for the horizontal components of the mean wind and for the vertical momentum flux. The momentum equation reads:

$$K_\tau \frac{d^2 \mathbf{V}}{dz^2} - \imath f \left(\mathbf{V} - \mathbf{V}_g \right) = 0 \;, \tag{S3.14}$$

whose solution is:

$$\mathbf{V} = a_1 \exp(b_1 z) + a_2 \exp(b_2 z) + \mathbf{V}_g \,, \tag{S3.15}$$

where, being $b_i^2 = \imath f/K_\tau$:

$$b_{1,2} = \mp(1+\imath)\left(\frac{f}{2K_\tau}\right)^{1/2} . \tag{S3.16}$$

From (3.69), $a_2 = 0$. Thus:

$$\mathbf{V} = a_1 \exp\left(-(f/2K_\tau)^{1/2} z\right) \left[\cos\left((f/2K_\tau)^{1/2} z\right) - \imath \sin\left((f/2K_\tau)^{1/2} z\right)\right] + \mathbf{V}_g \,, \tag{S3.17}$$

and for (3.68):

$$a_1 = -\mathbf{V}_g \,. \tag{S3.18}$$

The two components of the velocity result:

$$U = U_g\left[1 - \exp\left(-(f/2K_\tau)^{1/2} z\right)\cos\left((f/2K_\tau)^{1/2} z\right)\right] - V_g \exp\left(-(f/2K_\tau)^{1/2} z\right)\sin\left((f/2K_\tau)^{1/2} z\right) \,, \tag{S3.19}$$

$$V = V_g\left[1 - \exp\left(-(f/2K_\tau)^{1/2} z\right)\cos\left((f/2K_\tau)^{1/2} z\right)\right] + U_g \exp\left(-(f/2K_\tau)^{1/2} z\right)\sin\left((f/2K_\tau)^{1/2} z\right) \,. \tag{S3.20}$$

At the ground:

$$\frac{dU}{dz}\Big|_{z=0} = (f/2K_\tau)^{1/2}(U_g - V_g) \,, \tag{S3.21}$$

$$\frac{dV}{dz}\Big|_{z=0} = (f/2K_\tau)^{1/2}(U_g + V_g) \,. \tag{S3.22}$$

Orienting the coordinate axes such that $V_g = 0$ it results:

$$\frac{dU}{dz}\Big|_{z=0} = \frac{dV}{dz}\Big|_{z=0} = U_g(f/2K_\tau)^{1/2} \,, \tag{S3.23}$$

i.e. the mean velocity near the ground is rotated of $\pi/4$ (anticlockwise in the Northern hemisphere) with respect to the geostrophic wind.

From the flux-gradient relationship (2.80) the friction velocity is estimated:

$$u_*^2 = (K_\tau f)^{1/2}(U_g^2 + V_g^2)^{1/2} = (K_\tau f)^{1/2}|\mathbf{V}_g| \,. \tag{S3.24}$$

The wind attains the geostrophic velocity, from (S3.19) or (S3.20), at $h = \beta(f/2K_\tau)^{1/2}$. On the other hand, from the outer scaling argument, we obtain (3.16). So an expression for the eddy diffusion coefficient results:

$$K_\tau = \frac{\alpha^2 u_*^2}{2\beta^2 f} \tag{S3.25}$$

3.5 The integral of (3.110) results:

$$\kappa \frac{\Theta(z_2) - \Theta(z_1)}{-\theta_*} = \ln\frac{z_2}{z_1} + \alpha(\zeta_2 - \zeta_1)$$
$$+\beta_1 I_1(\zeta_2, \zeta_1) + (\beta_2 + \alpha\beta_1)\, I_2(\zeta_2, \zeta_1) + \alpha\beta_2\, I_3(\zeta_2, \zeta_1)\,, \tag{S3.26}$$

being

$$I_1(\zeta_2, \zeta_1) = \left[\frac{\ln(1+\beta_3\zeta)}{\beta_3}\right]_{\zeta_1}^{\zeta_2}, \tag{S3.27}$$

$$I_2(\zeta_2, \zeta_1) = \left[\frac{\zeta}{\beta_3} - \frac{\ln(1+\beta_3\zeta)}{\beta_3^2}\right]_{\zeta_1}^{\zeta_2}, \tag{S3.28}$$

$$I_3(\zeta_2, \zeta_1) = \left[\frac{\zeta^2}{2\beta_3} - \frac{1}{\beta_3}\left(\frac{\zeta}{\beta_3} - \frac{\ln(1+\beta_3\zeta)}{\beta_3^2}\right)\right]_{\zeta_1}^{\zeta_2}. \tag{S3.29}$$

Remember that in this formula $\zeta = z/\Lambda$ and Λ is assumed to be constant in the layer z_1, z_2.

3.6 Equations (3.115) and (3.116) are made nondimensional by defining $\mathbf{x} = \mathbf{X}/u_*^2$, $y = \langle\theta w\rangle/\langle\theta w\rangle_0$, $\mathbf{s}_V = L/u_*\, \mathrm{d}\mathbf{V}/\,\mathrm{d}z$, $s_\theta = -L/\theta_*\, \mathrm{d}\Theta/\,\mathrm{d}z$, where $u_*^2 = |\mathbf{X}(z=0)|$. Moreover $\xi = z/h$. Remember that $\theta_* < 0$.

Equation (3.115) becomes:

$$\frac{\mathrm{d}^2 y}{\mathrm{d}\xi^2} = 0\,, \tag{S3.30}$$

and (3.116) reads:

$$-\imath\frac{fh^2}{u_* L}s_V + \frac{\mathrm{d}^2 x}{\mathrm{d}\xi^2} = 0\,. \tag{S3.31}$$

The last equation suggests that the unique parameter characterising the solution is

$$\gamma_c^2 = \frac{fh^2}{u_* L}\,, \tag{S3.32}$$

but the equations are still unclosed.

From the definitions the following expressions for R_f and R_g can be obtained:

$$R_f \equiv \frac{g/\theta_{00}\langle\theta w\rangle}{\langle uw\rangle\, dU/\,dz + \langle vw\rangle\, dV/\,dz} = -\frac{g/\theta_{00}\langle\theta w\rangle}{\Re[\mathbf{X}^*\, d\mathbf{V}/\,dz]} = \frac{1}{\kappa}\frac{\mathbf{y}}{\Re[\mathbf{x}^*\mathbf{s}_V]}\,, \tag{S3.33}$$

($\mathbf{X}^*$ is the complex conjugate of $\mathbf{X}$),

$$R_g \equiv \frac{g/\theta_{00}\, d\Theta/\,dz}{(\,dU/\,dz)^2 + (\,dV/\,dz)^2} = \frac{g/\theta_{00}\, d\Theta/\,dz}{|\,d\mathbf{V}/\,dz|^2} = \frac{1}{\kappa}\frac{s_\theta}{|\mathbf{s}_V|^2} \tag{S3.34}$$

The closure of the model arises from the assumption that the two Richardson numbers are constant and equal to R_{fc} and R_{gc}.

The boundary conditions for Eqs. (S3.30) and (S3.31) are:

$$y = 1\,,\ |\mathbf{x}| = 1 \text{ for } \xi = 0 \tag{S3.35}$$

$$y = 0\,,\ |\mathbf{x}| = 0 \text{ for } \xi = 1 \tag{S3.36}$$

$$\mathbf{V} = 0 \text{ for } \xi = 0 \tag{S3.37}$$

$$\mathbf{V} = \mathbf{V}_g \text{ for } \xi = 1 \tag{S3.38}$$

Equation (S3.30), remembering the boundary conditions above, has the linear solution:

$$y = (1 - \xi)\,, \tag{S3.39}$$

i.e. (3.121).

Equation (S3.33) is rewritten using (S3.39):

$$\Re[\mathbf{x}^*\mathbf{s}_V] = \frac{1-\xi}{\kappa R_{fc}}\,, \tag{S3.40}$$

or

$$\mathbf{x}^*\mathbf{s}_V = \frac{1-\xi}{\kappa R_{fc}}\,,\ \Im[\mathbf{x}^*\mathbf{s}_V] = 0\,. \tag{S3.41}$$

By multiplying (S3.31) by $\mathbf{x}^*$ and substituting with (S3.41) we get:

$$\mathbf{x}^*\frac{d^2\mathbf{x}}{d\xi^2} - \imath\frac{\gamma_c^2}{\kappa R_{fc}}(1-\xi) = 0\,. \tag{S3.42}$$

We are looking for a solution in the form $\mathbf{x} = (1-\xi)^{(a+\imath b)}$. By substituting in Eq. (S3.42) we get:

$$(a+\imath b)(a-1+\imath b)(1-\xi)^{2a-2} - \imath \frac{\gamma_c^2}{\kappa R_{\mathrm{fc}}}(1-\xi) = 0 \,. \tag{S3.43}$$

In order to verify the identity, the exponents of $(1-\xi)$ must be equal: $2a-2=1$, and thus $a = 3/2$. Using this value for a, and putting equal the real and imaginary parts of the multiplicative coefficients, it results:

$$\frac{3}{4} - b^2 = 0 \,,\, 2b = \frac{\gamma_{\mathrm{c}}^2}{\kappa R_{\mathrm{fc}}} \,. \tag{S3.44}$$

Thus, from the first relationship, $b = \sqrt{3}/2$, and using this value in the second relationship:

$$\frac{\gamma_{\mathrm{c}}^2}{\kappa R_{\mathrm{fc}}} = \sqrt{3} \,. \tag{S3.45}$$

This equation allows to write the height h in terms of the other model parameters: from Eq. (S3.32) we get the relation (3.124).

The solution of (S3.42) reads:

$$\begin{aligned} \mathbf{x} = (1-\xi)^{(3+\imath\sqrt{3})/2} \equiv \\ \equiv (1-\xi)^{3/2} \exp\left(\imath \frac{\sqrt{3}}{2} \ln(1-\xi)\right) \equiv \\ \equiv (1-\xi)^{3/2} \left[\cos\left(\frac{\sqrt{3}}{2} \ln(1-\xi)\right) + \imath \sin\left(\frac{\sqrt{3}}{2} \ln(1-\xi)\right)\right] \,, \end{aligned} \tag{S3.46}$$

the same as (3.122).

Using (S3.46) into (S3.41) we obtain the expression for the vertical gradient of the mean velocity:

$$\mathbf{s}_{\mathrm{V}} = \frac{1}{\kappa R_{\mathrm{fc}}}(1-\xi)^{(-1+\imath\sqrt{3})/2} \,, \tag{S3.47}$$

which is the same as (3.125). Verify that the solution for $\mathbf{x}$ and $\mathbf{s}_{\mathrm{V}}$ satisfies the condition $\Im[\mathbf{x}^* \mathbf{s}_{\mathrm{V}}] = 0$.

By substituting (S3.47) into (S3.34), the expression for the vertical gradient of the mean temperature can be obtained:

$$s_\theta = \frac{R_{\mathrm{gc}}}{\kappa R_{\mathrm{fc}}^2}(1-\xi)^{-1} \,, \tag{S3.48}$$

the same as (3.126).

The vertical profile of mean wind is obtained by integrating (S3.47). Remembering (S3.37) we get:

$$\mathbf{V}(z) = \frac{u_*}{L\kappa R_{\mathrm{fc}}} \int_0^z \left(1 - \frac{z'}{h}\right)^{(-1+\imath\sqrt{3})/2} \mathrm{d}z' \,, \tag{S3.49}$$

and putting $t = 1 - z'/h$ we obtain:

$$\mathbf{V}(z) = -\frac{u_* h}{L\kappa R_{\mathrm{fc}}} \int_1^{1-z/h} t^{(-1+\imath\sqrt{3})/2} \, \mathrm{d}t =$$

$$= \frac{u_* h}{L\kappa R_{\mathrm{fc}}} \frac{1-\imath\sqrt{3}}{2} \left[1 - (1 - z/h)^{(1+\imath\sqrt{3})/2}\right] . \tag{S3.50}$$

According to the condition (S3.38), i.e. $\mathbf{V}(h) = \mathbf{V}_{\mathrm{g}}$, an expression for $\mathbf{V}_{\mathrm{g}}$ can be obtained as function of the model parameters:

$$\mathbf{V}_{\mathrm{g}} = \frac{u_* h}{L\kappa R_{\mathrm{fc}}} \frac{1-\imath\sqrt{3}}{2} = \frac{u_* h}{L\kappa R_{\mathrm{fc}}} \mathrm{e}^{-\frac{i\pi}{3}} . \tag{S3.51}$$

The modulus of the geostrophic wind results

$$|\mathbf{V}_{\mathrm{g}}| = \frac{u_* h}{L\kappa R_{\mathrm{fc}}} \,, \tag{S3.52}$$

or

$$|\mathbf{V}_{\mathrm{g}}| = \left(\frac{\sqrt{3}}{\kappa R_{\mathrm{fc}}}\right) \frac{u_*^{3/2}}{L^{1/2} f^{1/2}} \,, \tag{S3.53}$$

and its components are

$$U_{\mathrm{g}} = |\mathbf{V}_{\mathrm{g}}| \cos\left(-\frac{\pi}{3}\right) \,, V_{\mathrm{g}} = |\mathbf{V}_{\mathrm{g}}| \sin\left(-\frac{\pi}{3}\right) . \tag{S3.54}$$

For the vertical profile of mean velocity an Ekman-like spiral is obtained:

$$\frac{\mathbf{V}_{\mathrm{g}} - \mathbf{V}}{|\mathbf{V}_{\mathrm{g}}|} = \mathrm{e}^{\frac{i\pi}{3}} \left(1 - \frac{z}{h}\right)^{\frac{1}{2}(1+i\sqrt{3})} \,, \tag{S3.55}$$

or

$$\frac{U_{\mathrm{g}} - U}{|\mathbf{V}_{\mathrm{g}}|} = \left(1 - \frac{z}{h}\right)^{1/2} \cos\left[\frac{\sqrt{3}}{2} \ln\left(1 - \frac{z}{h}\right) - \frac{\pi}{3}\right] \,, \tag{S3.56}$$

$$\frac{V_{\mathrm{g}} - V}{|\mathbf{V}_{\mathrm{g}}|} = \left(1 - \frac{z}{h}\right)^{1/2} \sin\left[\frac{\sqrt{3}}{2} \ln\left(1 - \frac{z}{h}\right) - \frac{\pi}{3}\right] . \tag{S3.57}$$

Sometimes it is useful the modulus of the mean wind $|\mathbf{V}(z)| \equiv \sqrt{\mathbf{V}^*(z)\mathbf{V}(z)}$:

$$|\mathbf{V}(z)| = |\mathbf{V}_g| \left[2 - \frac{z}{h} - 2\left(1-\frac{z}{h}\right)^{\frac{1}{2}} \cos\left(\frac{\sqrt{3}}{2} \ln\left(1-\frac{z}{h}\right)\right)\right]^{1/2} . \tag{S3.58}$$

For $z/h \ll 1$ the expansion of (S3.50) gives:

$$\mathbf{V}(z) \equiv U(z) + \imath V(z) = \frac{u_* z}{L\kappa R_{\mathrm{fc}}} . \tag{S3.59}$$

The vertical profile of mean temperature is obtained by integrating (S3.48):

$$\Theta(z) = \Theta(0) - \frac{\theta_*}{L} \frac{R_{\mathrm{gc}}}{\kappa R_{\mathrm{fc}}^2} \int_0^z \left(1 - \frac{z'}{h}\right)^{-1} \mathrm{d}z' , \tag{S3.60}$$

and the temperature profile reads:

$$\Theta(z) = \Theta(0) + \frac{\theta_* h}{L} \frac{R_{\mathrm{gc}}}{\kappa R_{\mathrm{fc}}^2} \ln\left(1 - \frac{z}{h}\right) , z < h . \tag{S3.61}$$

For $z/h \ll 1$ the series expansion of (S3.61), being $\ln(1-x) \simeq -x$, gives:

$$\Theta(z) = \Theta(0) - \frac{\theta_* R_{\mathrm{gc}} z}{\kappa R_{\mathrm{fc}}^2 L} . \tag{S3.62}$$

The eddy diffusion coefficients consistent with this model are obtained using the flux-gradient relationship. As far the momentum is concerned:

$$-\mathbf{X} = -K_\tau \frac{\mathrm{d}\mathbf{V}}{\mathrm{d}z} , \tag{S3.63}$$

and using (S3.46) and (S3.47) we obtain:

$$K_\tau = u_* \kappa L R_{\mathrm{fc}} \left(1 - \xi\right)^2 . \tag{S3.64}$$

Similarly, as far temperature is concerned:

$$y\, u_*\, \theta_* = -K_{\mathrm{Q}} \frac{\mathrm{d}\Theta}{\mathrm{d}z} = K_{\mathrm{Q}} s_\theta \frac{\theta_*}{L} , \tag{S3.65}$$

from which, using (3.121) and (S3.48) we get:

$$K_{\mathrm{Q}} = u_* \kappa L \frac{R_{\mathrm{fc}}^2}{R_{\mathrm{gc}}} \left(1 - \xi\right)^2 . \tag{S3.66}$$

3.7 A comparison between data and model needs an evaluation of the SBL depth and of the Obukhov length at the surface. Using tower data a direct estimate of h is not available. Here the following procedure has been used:

- from (3.124), with the coefficient $\gamma_c = 0.4$, it results:

$$\frac{h}{L} = \frac{\gamma_c}{f^{1/2}} \left(-\frac{\theta_{00} \langle w\theta \rangle}{g\kappa u_*^2} \right)^{1/2} , \tag{S3.67}$$

  i.e., the RHS is essentially function of the surface fluxes;
- tSBL observations with a restricted range of surface fluxes are selected (in the figure, $0.1 < u_* < 0.3\,\mathrm{m}^{-1}$, $0.06 < -\theta_* < 0.1\,\mathrm{K}$: note that the choice is arbitrary, and is dictated by the availability of data);
- for each observation, L can be derived (from surface data extrapolated from the observed fluxes, for instance) and thus h. With this choice of parameters, it results $h/L \approx 2.74$;
- the measurement height is scaled with the value of h, and each observation is reported in the figure at the corresponding nondimensional height;
- the surface layer profiles are computed by choosing a specific value of h, and using the value of L derived from the previous estimate. In the figure, two values of h are used, 50 and 100 m, which corresponds to the most populated range of depths, computed from (3.124), for the selected dataset.

Exercises of Chap. 4

4.1 Let we consider the domain represented in Fig. S4.1 and take the average over the area A (which includes obstacles) of the derivative of the scalar quantity a with respect to x_1:

$$\left\langle \frac{\partial a}{\partial x_1} \right\rangle = \frac{1}{A} \int\int_A \frac{\partial a}{\partial x_1}\, \mathrm{d}x_1\, \mathrm{d}x_2 = \frac{1}{A} \left[\int_{C_0} a\, \mathrm{d}x_2 - \sum_i \int_{C_i} a\, \mathrm{d}x_2 \right] \tag{S4.1}$$

by the Green theorem: the average of the derivative is equal to the integral along the path C_0 bounding the averaging area, plus the sum of the integrals along the paths bounding the internal obstacles. Note that in the case of a rectangular domain:

$$\int_{C_0} a\, \mathrm{d}x_2 =$$

$$\int_{\xi_2-\Delta x_2/2}^{\xi_2+\Delta x_2/2} a(\xi_1 + \Delta x_1/2, x_2)\, \mathrm{d}x_2 + \int_{\xi_2+\Delta x_2/2}^{\xi_2-\Delta x_2/2} a(\xi_1 - \Delta x_1/2, x_2)\, \mathrm{d}x_2 \tag{S4.2}$$

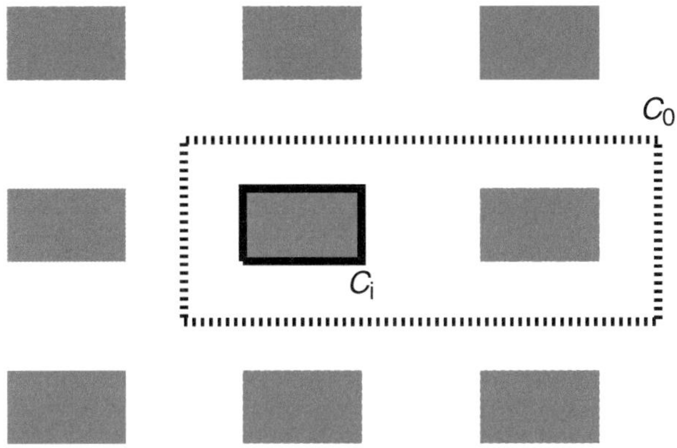

Fig. S4.1 Scheme for the area averaging. The path C_0 surrounding the averaging area (*dashed line*) does not encounter obstacles, the other paths C_i surround each obstacle

On the other hand

$$\frac{\partial \langle a \rangle}{\partial \xi_1} = \frac{1}{A} \int_{\xi_2-\Delta x_2/2}^{\xi_2+\Delta x_2/2} \mathrm{d}x_2$$

$$\left[\int_{\xi_2-\Delta x_2/2}^{\xi_2+\Delta x_2/2} \frac{\partial a}{\partial \xi_1} \mathrm{d}x_1 + a(\xi_2 + \Delta x_2/2, x_2) - a(\xi_2 - \Delta x_2/2, x_2) \right] \quad \text{(S4.3)}$$

Because a is not a function of ξ_1 the first term into the squared parenthesis is zero, Eq. (S4.3) reduces to

$$\frac{\partial \langle a \rangle}{\partial \xi_1} = \frac{1}{A} \int_{\xi_2-\Delta x_2/2}^{\xi_2+\Delta x_2/2} \mathrm{d}x_2 \left[a(\xi_1 + \Delta x_1/2, x_2) - a(\xi_1 - \Delta x_1/2, x_2) \right] \quad \text{(S4.4)}$$

thus the derivative and the area averaging commutes only if the integrals on the inner paths are zero, i.e. if

$$\sum_i \int_{C_i} a \, \mathrm{d}x_2 = 0 \quad \text{(S4.5)}$$

Let consider a few cases (this discussion follows Raupach and Shaw 1982). If the mean pressure is considered, with the mean velocity positive in x direction (see Fig. S4.2):

$$\left\langle \frac{\partial P}{\partial x} \right\rangle = \left\langle \frac{\partial \langle P \rangle}{\partial x} \right\rangle + \left\langle \frac{\partial P'}{\partial x} \right\rangle = \frac{\partial \langle P \rangle}{\partial \xi} + \left\langle \frac{\partial P'}{\partial x} \right\rangle \quad \text{(S4.6)}$$

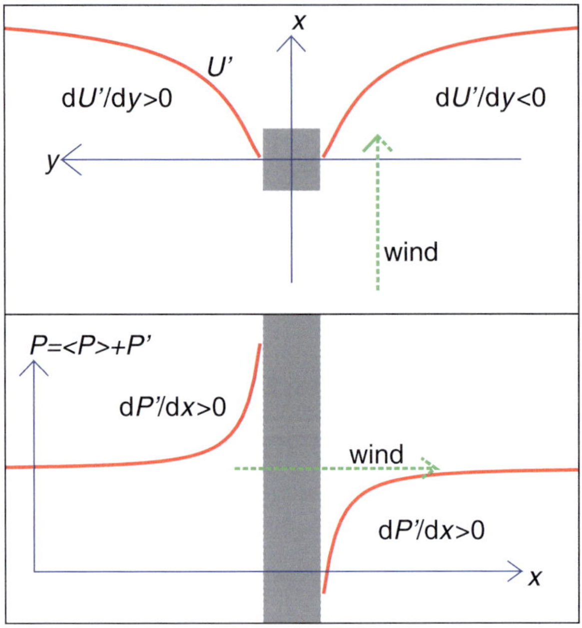

Fig. S4.2 (**a**) schematic field of wind near the sides of a bluff body parallel to mean flow. (**b**) schematic pressure field about a bluff body facing the wind

The area-averaged derivative of $\langle P\rangle$ commutes because it is constant inside A (and thus along the inner paths). But $\partial P'/\partial x > 0$ so the area-averaged value of the derivative is greater than zero too.

A second example concerns velocity. In this case averaging commutes with derivation because velocity is zero on the paths contouring obstacles, so $U_i' = U_i - \langle U_i\rangle$ is constant on C_i and thus $\langle \partial U_i'/\partial x\rangle = 0$. This result holds for all the moments of velocity. As third example, let consider the second derivatives of velocity: see Fig. S4.2.

In this case $\partial U/\partial y$ is not constant along C_i, so that taking the average of the second order derivative gives

$$\left\langle \frac{\partial^2 U}{\partial y^2} \right\rangle = \frac{\partial^2 \langle U\rangle}{\partial \xi^2} + \left\langle \frac{\partial^2 U'}{\partial y^2} \right\rangle \tag{S4.7}$$

Thus, from the Reynolds equations (2.62) for a neutral flow and neglecting the Coriolis term, taking the area average it results

$$\frac{\partial\langle U_i\rangle}{\partial t} + \langle U_j\rangle \frac{\partial\langle U_i\rangle}{\partial \xi_j} =$$

$$-\frac{1}{\rho_{00}}\frac{\partial\langle P\rangle}{\partial \xi_i} + \nu\frac{\partial^2\langle U_i\rangle}{\partial \xi_j\partial \xi_j} + \frac{\partial \tau_{ij}}{\partial \xi_j} - \frac{1}{\rho_{00}}\left\langle \frac{\partial P}{\partial x_i}\right\rangle + \nu\left\langle \frac{\partial^2 U_i}{\partial x_j \partial x_j}\right\rangle, \tag{S4.8}$$

where τ_{ij} is the negative of the area-averaged tensor of the turbulent fluxes and the last two terms on the RHS represent the form drag and the viscous drag due to the presence of obstacles.

4.2 The mean wind profile for $z > h_c$ is given by (3.33) and for $z \leq h_c$ by (4.63). The profiles match at $z = h_c$ if we write:

$$u(h_c) = \frac{u_*}{\kappa} \ln \frac{h_c - z_d}{z_0} , \tag{S4.9}$$

from the profile (3.33) and thus

$$u(z) = \frac{u_*}{\kappa} \ln \frac{h_c - z_d}{z_0} \exp\left[a\left(\frac{z}{h_c} - 1\right)\right], \text{ for } z \leq h_c . \tag{S4.10}$$

To obtain a smooth profile, we need also to match the derivatives at $z = h_c$. For $z > h_c$ the derivative reads:

$$\frac{dU}{dz} = \frac{u_*}{\kappa(z - z_d)} , \tag{S4.11}$$

while for $z \leq h_c$:

$$\frac{dU}{dz} = \frac{u_* a}{\kappa h_c} \ln \frac{h_c - z_d}{z_0} \exp\left[a\left(\frac{z}{h_c} - 1\right)\right] . \tag{S4.12}$$

The two derivatives are equal in $z = h_c$ if

$$\frac{1}{h_c - z_d} = \frac{a}{h_c} \ln \frac{h_c - z_d}{z_0} , \tag{S4.13}$$

which is a nonlinear equation in the variable $\chi = (h_c - z_d)/z_0$, so that the value of z_d can be used to match the derivatives, as the characteristics of the canopy (height, density, roughness) are given.

4.3 To compute the profile variations as the canopy features (density, height) change, the mean profiles are given by (3.33) and (S4.10), the momentum flux in the canopy is given by (4.65) and above is constant $\tau_{13}(h_c) \equiv u_*^2$. These profiles are normalized to the value at a fixed height, much greater than the building height: at such height we assume that the wind velocity takes its mean value not affected by the surface changes. For the exercise, take $z_r = 10\, h_{c1}$.

We prescribe the values of h_c and L_c that characterize the two canopies, and need some simple parameterizations to estimate the values of z_0, z_d and l_c (and thus a): we use here $l_c = 0.1\, h_c$ then compute a and $z_0 = h_c/30\, a^{0.1}$ (the weak dependence on the canopy parameters leads to a roughness that increases as the canopy density increases, with the same obstacle height). Then, the displacement height z_d is derived from (S4.13).

Table S4.1 Canopy parameters for Exercise 4.3

C1	$h_c = 1$	$L_c = 2$	$a = 2.92$	$z_d = 0.80h_c$
C2	$h_c = 1$	$L_c = 1$	$a = 3.86$	$z_d = 0.83h_c$
C3	$h_c = 4$	$L_c = 8$	$a = 2.92$	$z_d = 0.80h_c$

All the lengths made nondimensional with the C1 height

The parameters used in the computations for the three canopies are reported in Table S4.1. Case 1 is obtained looking at a transition from C1 to C2, case 2 at a transition from C1 to C3.

Exercises of Chap. 5

5.1 In Eqs. (5.15)–(5.17) the integral of the position in z:

$$z(t) = \int_0^t w(t_1)\,\mathrm{d}t_1 \tag{S5.1}$$

appears together with the integrals of the velocity components. Equation (5.15) is written as the sum of the double integral of three terms: the first one is proportional to

$$\begin{aligned}\langle z(t_1)z(t_2)\rangle &= \int_0^{t_1} \mathrm{d}t_1' \int_0^{t_2} \mathrm{d}t_2' R_{33}(t_1' - t_2') = \\ &= \int_0^{t_1} (t_1 - t')R_{33}(t')\,\mathrm{d}t' + \int_0^{t_2} (t_2 - t')R_{33}(t')\,\mathrm{d}t' + \\ &\quad - \int_0^{|t_1-t_2|} (|t_1 - t_2| - t')R_{33}(t')\,\mathrm{d}t'\,,\end{aligned} \tag{S5.2}$$

the second one is proportional to

$$\begin{aligned}&\langle z(t_1)u(t_2)\rangle + \langle z(t_2)u(t_1)\rangle \\ &= \int_0^{t_1} R_{13}(t_2 - t')\,\mathrm{d}t' + \int_0^{t_2} R_{31}(t_1 - t')\,\mathrm{d}t' \\ &= \int_0^{t_1} R_{13}(t')\,\mathrm{d}t' + \int_0^{t_2} R_{31}(t')\,\mathrm{d}t' - \int_0^{|t_1-t_2|} \left[R_{13}(t') - R_{31}(t')\right]\mathrm{d}t' \\ &= \int_0^{t_1} R_{13}(t')\,\mathrm{d}t' + \int_0^{t_2} R_{13}(t')\,\mathrm{d}t\,,\end{aligned} \tag{S5.3}$$

because $R_{13} = R_{31}$, and the third one is given by

$$\langle u(t_1)u(t_2)\rangle = R_{11}(t_1 - t_2) \ . \tag{S5.4}$$

Thus (5.15) gives

$$\begin{aligned} \langle x^2\rangle(t) = \frac{\Gamma^2}{3} \int_0^t \left(2t^3 - 3t^2 s + s^3\right) R_{33}(s)\,\mathrm{d}s \\ +2\Gamma \int_0^t t(t-t)R_{13}(s)\,\mathrm{d}s \\ +2\int_0^t (t-s)R_{11}(s)\,\mathrm{d}s \ . \end{aligned} \tag{S5.5}$$

Similar integrals appear in (5.16), that becomes:

$$\langle xz\rangle(t) = \Gamma t \int_0^t (t-s)R_{33}(s)\,\mathrm{d}s + 2\int_0^t (t-s)R_{13}(s)\,\mathrm{d}s \ , \tag{S5.6}$$

whereas for $\langle y^2\rangle$ and for $\langle z^2\rangle$ the solution is (5.6).

Using the exponential form (2.112), written here $R_{ij}(s) = \langle u_i u_j\rangle \exp(-s/T_{ij})$, the previous integrals may be done explicitly. Thus (S5.5) reads:

$$\langle x^2\rangle(t) = I_1 + I_2 + I_3 \ , \tag{S5.7}$$

where:

$$\begin{aligned} I_1(t) = {} & \frac{\langle w^2\rangle \Gamma^3 T_{33}}{3} \{2t^3 \left[1 - \exp(-t/T_{33})\right] \\ & - 3t^2 T_{33} \left[1 - (1 + t/T_{33})\exp(-t/T_{33})\right] \\ & + T_{33}^3 \left[6 - \left(6 + 6t/T_{33} + 3t^2/T_{33}^2 + t^3/T_{33}^3\right)\exp(-t/T_{33})\right]\} \ , \end{aligned} \tag{S5.8}$$

$$\begin{aligned} I_2(t) = {} & 2\langle uw\rangle \Gamma T_{13} t \\ & \{t\left[1 - \exp(-t/T_{13})\right] - 2T_{13}\left[1 - (1 + t/T_{13})\exp(-t/T_{13})\right]\} \ , \end{aligned} \tag{S5.9}$$

$$I_3(t) = 2\langle u^2\rangle T_{11}^2 \left[t/T_{11} - 1 + \exp(-t/T_{11})\right] \ . \tag{S5.10}$$

Similarly, (S5.6) reads:

$$\begin{aligned} \langle xz\rangle(t) = {} & 2\langle w^2\rangle \Gamma T_{33}^2 t \left[t/T_{33} - 1 + \exp(-t/T_{33})\right] \\ & + 4\langle uw\rangle T_{13}^2 \left[t/T_{13} - 1 + \exp(-t/T_{13})\right] \ . \end{aligned} \tag{S5.11}$$

5.2 The integral of (5.52), with the conditions $x_i(0) = 0$, $a_i \equiv U_i = 0$, $b_{ij} = 2D\delta_{ij}$ as in the text, reads:

$$x_i(t) = \sqrt{2D} \int_0^t \mathrm{d}W(t') \,, \tag{S5.12}$$

and taking the average:

$$\langle x_i \rangle (t) = 0 \,, \tag{S5.13}$$

while taking the square and averaging:

$$\begin{aligned} \langle x_i^2 \rangle &= 2D \langle \int_0^t \mathrm{d}W(t_1) \int_0^t \mathrm{d}W(t_2) \rangle \\ &= 2D \int_0^t \int_0^t \langle \mathrm{d}W(t_1)\, \mathrm{d}W(t_2) \rangle = 2Dt \,, \end{aligned} \tag{S5.14}$$

thanks to (5.49).

5.3 The stochastic equation (5.82) can be integrated (Gardiner 1990, p. 106) as follows. Putting

$$w = u \exp(t/T_{\mathrm{uu}}) \,, \tag{S5.15}$$

then

$$\mathrm{d}w = \mathrm{d}u \exp(t/T_{\mathrm{uu}}) + \frac{u}{T_{\mathrm{uu}}} \exp(t/T_{\mathrm{uu}})\, \mathrm{d}t = \left(\frac{2\langle u^2 \rangle}{T_{\mathrm{uu}}} \right)^{1/2} \exp(t/T_{\mathrm{uu}})\, \mathrm{d}W \,. \tag{S5.16}$$

By integrating:

$$w(t) - w(0) = \left(\frac{2\langle u^2 \rangle}{T_{\mathrm{uu}}} \right)^{1/2} \int_0^t \exp(s/T_{\mathrm{uu}})\, \mathrm{d}W(s) \tag{S5.17}$$

and substituting with (S5.15) the expression (5.83) results.

The average velocity is given by (5.84) taking the average over the particles that leave the source.

The covariance $\langle u(t)u(s) \rangle_{\mathrm{S}}$ reads:

$$\begin{aligned} \langle u(t)u(s) \rangle_{\mathrm{S}} &= \langle u(0)^2 \rangle_{\mathrm{S}} \exp[-(t+s)/T_{\mathrm{uu}}] + \\ &+ \left(\frac{2\langle u^2 \rangle}{T_{\mathrm{uu}}} \right) \int_0^t \int_0^s \exp[-(t-t')/T_{\mathrm{uu}}] \exp[-(s-s')/T_{\mathrm{uu}}] \langle \mathrm{d}W(t')\, \mathrm{d}W(s') \rangle \,. \end{aligned} \tag{S5.18}$$

Let consider the case of $t > s$. Using (5.49) the second term of the RHS of (S5.18) may be rewritten as follows:

$$\int_0^t \int_0^s \exp[-(t-t')/T_{\mathrm{uu}}] \exp[-(s-s')/T_{\mathrm{uu}}] \langle \mathrm{d}W(t') \, \mathrm{d}W(s') \rangle$$
$$= \int_0^s \int_0^s \exp[-(t-t')/T_{\mathrm{uu}}] \exp[-(s-s')/T_{\mathrm{uu}}] \langle \mathrm{d}W(t') \, \mathrm{d}W(s') \rangle$$
$$+ \int_s^t \int_0^s \exp[-(t-t')/T_{\mathrm{uu}}] \exp[-(s-s')/T_{\mathrm{uu}}] \langle \mathrm{d}W(t') \, \mathrm{d}W(s') \rangle \, , \quad \text{(S5.19)}$$

where the second term in RHS is zero because t' (being $s < t' < t$) never superimposes with s' (being $0 < s' < s$) and thus in $\delta(t' - s')$ it always occurs $t' \neq s'$.

Then integrating in t':

$$\begin{aligned}
\langle u(t)u(s) \rangle_{\mathrm{S}} & \\
&= \langle u(0)^2 \rangle_{\mathrm{S}} \exp[-(t+s)/T_{\mathrm{uu}}] \\
&+ \left(\frac{2\langle u^2 \rangle}{T_{\mathrm{uu}}} \right) \int_0^s \exp[-(t+s-2s')/T_{\mathrm{uu}}] \, \mathrm{d}s' \\
&= \langle u(0)^2 \rangle_{\mathrm{S}} \exp[-(t+s)/T_{\mathrm{uu}}] \\
&+ \left(\frac{2\langle u^2 \rangle}{T_{\mathrm{uu}}} \right) \exp[-(t+s)/T_{\mathrm{uu}}] \frac{T_{\mathrm{uu}}}{2} [\exp(2s/T_{\mathrm{uu}}) - 1] = \\
&= (\langle u(0)^2 \rangle_{\mathrm{S}} - \langle u^2 \rangle) \exp[-(t+s)/T_{\mathrm{uu}}] \\
&+ \langle u^2 \rangle \exp[-(t-s)/T_{\mathrm{uu}}] \qquad \text{(S5.20)}
\end{aligned}$$

If $s > t$ the same procedure must be made, integrating in $\mathrm{d}s'$ and reducing to an integral in $\mathrm{d}t'$ from 0 to t, leading to an expression like (S5.20) with s instead of t and viceversa. The general form (5.85) for the covariance then results.

5.4 Flow and particle velocity component u and v (here we drop the index for brevity, and the tilde over u) are expressed by Fourier transform:

$$(u, v) = \int_{-\infty}^{\infty} (\hat{u}, \hat{v}) \exp(\imath \omega t) \, \mathrm{d}\omega \, . \qquad \text{(S5.21)}$$

From (5.102) it results:

$$\hat{v} = (i\omega T_{\mathrm{p}} + 1)^{-1} \hat{u} \, , \qquad \text{(S5.22)}$$

and transforming:

$$v(t) = \int_{-\infty}^{+\infty} \frac{1 - \imath\omega T_{\mathrm{p}}}{1 + \omega^2 T_{\mathrm{p}}^2} \hat{u} \exp(\imath\omega t)\, \mathrm{d}\omega \; . \tag{S5.23}$$

The convolution of the two functions $f(t)$ and $g(t)$ is:

$$[f \circ g](t) = \frac{1}{2\pi} \int_{-\infty}^{+\infty} f(\tau) g(t-\tau)\, \mathrm{d}\tau = \frac{1}{2\pi} \int_{-\infty}^{+\infty} f(t-\tau) g(\tau)\, \mathrm{d}\tau \; , \tag{S5.24}$$

and remember (see, for instance, Gradshteyn and Ryzhik 1980, p. 1147) that:

$$\hat{f}(\omega)\hat{g}(\omega) = \frac{1}{2\pi} \int_{-\infty}^{\infty} [f \circ g](t) \exp(-\imath\omega t)\, \mathrm{d}t \; , \tag{S5.25}$$

i.e., the product of the transformed functions is equal to the Fourier transform of the convolution. From (S5.23) we recognize that the integral is the anti-transform of the product of two Fourier transforms, thus it is sufficient to write the anti-transform of the first factor under the integral to get the required expression. Then:

$$\int_{-\infty}^{+\infty} \frac{1 - \imath\omega T_{\mathrm{p}}}{1 + \omega^2 T_{\mathrm{p}}^2} \exp(\imath\omega t)\, \mathrm{d}\omega = \left[\frac{\pi}{T_{\mathrm{p}}} \exp(-|t|/T_{\mathrm{p}}) - \pi \frac{\mathrm{d}}{\mathrm{d}t} \exp(-|t|/T_{\mathrm{p}}) \right] , \tag{S5.26}$$

so that:

$$\begin{aligned} v(t) &= \left[\left(\frac{\pi}{T_{\mathrm{p}}} \left(1 - \frac{t}{|t|} \right) \exp(-|t|/T_{\mathrm{p}}) \right) \circ u(t) \right] \\ &= \frac{1}{T_{\mathrm{p}}} \int_{-\infty}^{t} \exp[-(t-s)/T_{\mathrm{p}}] u(s)\, \mathrm{d}s \end{aligned} \tag{S5.27}$$

5.5 The formal solution can be written starting from (S5.27). However, a simpler approach can be devised as follows.

Consider the time covariance of a velocity component. According to the definition (2.42) and substituting the ensemble average with the time average (2.40), it results:

$$R(\vartheta, t) = \langle u(\vartheta) u(\vartheta + t) \rangle = \frac{1}{T_{\mathrm{A}}} \int_{\vartheta - T_{\mathrm{A}}/2}^{\vartheta + T_{\mathrm{A}}/2} u(\tau) u(\tau + t)\, \mathrm{d}\tau \; , \tag{S5.28}$$

and in homogeneous steady conditions the value of ϑ is not relevant as well as the sign of the lag t. Put $\vartheta = 0$. Assume a long averaging time (note that dealing with measured time series, the averaging time cannot be longer than the total length of the time series, i.e. it is in fact finite) we can write approximately:

$$T_{\mathrm{A}} R(t) \approx \int_{-\infty}^{+\infty} u(\tau) u(\tau + t)\, \mathrm{d}\tau \equiv \int_{-\infty}^{+\infty} u(\tau - t) u(\tau)\, \mathrm{d}\tau \; . \tag{S5.29}$$

Consider now the definition of convolution (S5.24) and its Fourier transform (S5.25):

$$\hat{f}(\omega)\hat{g}(\omega) = \frac{1}{2\pi}\int_{-\infty}^{\infty} \exp(-\imath\omega t)\,\mathrm{d}t \frac{1}{2\pi}\int_{-\infty}^{+\infty} f(t-\tau)g(\tau)\,\mathrm{d}\tau\ , \tag{S5.30}$$

and put $f(t) = u(-t)$, $g(t) = u(t)$. Both f and g are real functions. For a real function:

$$\hat{f}^*(\omega) = \frac{1}{2\pi}\int_{-\infty}^{+\infty} f(t)\exp(+\imath\omega t)\,\mathrm{d}t = \frac{1}{2\pi}\int_{-\infty}^{+\infty} f(-t)\exp(-\imath\omega t)\,\mathrm{d}t\ , \tag{S5.31}$$

thus the Fourier transform of $f(t)$ is the complex conjugate of the Fourier transform of $g(t)$: $\hat{f(\omega)} = \hat{g}^*(\omega)$.

With this choice of the functions f and g, (S5.30) reads:

$$\begin{aligned}\hat{u}^*(\omega)\hat{u}(\omega) &= \frac{1}{2\pi}\int_{-\infty}^{\infty} \exp(-\imath\omega t)\,\mathrm{d}t \frac{1}{2\pi}\int_{-\infty}^{+\infty} u(\tau-t)u(\tau)\,\mathrm{d}\tau \\ &\approx \frac{T_\mathrm{A}}{(2\pi)^2}\int_{-\infty}^{\infty} \exp(-\imath\omega t)R(t)\,\mathrm{d}t\ ,\end{aligned} \tag{S5.32}$$

and, according to (2.55):

$$\hat{u}^*(\omega)\hat{u}(\omega) \approx \frac{T_\mathrm{A}}{4\pi}E(\omega)\ . \tag{S5.33}$$

Now, using (S5.22), the relation between the power spectrum for particles and that of the flow can be obtained:

$$\begin{aligned}E^{(p)}(\omega) &= \frac{4\pi}{T_\mathrm{A}}\hat{v}^*\hat{v} = \\ \frac{4\pi}{T_\mathrm{A}}\frac{(1+\imath\omega T_\mathrm{P})(1-\imath\omega T_\mathrm{P})}{(1+\omega^2T_\mathrm{P}^2)^2}\hat{u}^*\hat{u} &= \\ &\frac{E(\omega)}{1+\omega^2T_\mathrm{P}^2}\ .\end{aligned} \tag{S5.34}$$

The relation between the Lagrangian spectra can be used to derive the relation between the velocity variances. Using (2.57):

$$\langle v^2\rangle = \int_0^{\infty} \frac{E(\omega)}{1+\omega^2T_\mathrm{P}^2}\,\mathrm{d}\omega\ , \tag{S5.35}$$

and substituting with the expression of the Lagrangian spectrum (2.114), it results:

$$\langle v^2 \rangle = \frac{2\langle u^2 \rangle}{\pi} \int_0^{\infty} \frac{\mathrm{d}z}{[1 + (T_{\mathrm{P}}/T)^2 z^2]\,[1 + z^2]} , \tag{S5.36}$$

with $z = \omega T$. For the integral, see Gradshteyn and Ryzhik (1980, p. 67). It results:

$$\langle v^2 \rangle = \frac{\langle u^2 \rangle}{1 + (T_{\mathrm{P}}/T)} \tag{S5.37}$$

Exercises of Chap. 6

6.1 From the equations for the second-order moments, considering horizontal homogeneity, in the surface layer approximation ($V = 0$, $W = 0$), the following relationships are derived (Mellor and Yamada 1982).

$$\langle u^2 \rangle = \frac{E_{\mathrm{k}}}{3} + \frac{l_1}{E_{\mathrm{k}}^{1/2}} \left[-4\langle uw \rangle \frac{\mathrm{d}U}{\mathrm{d}z} - 2\frac{g}{\vartheta_{00}} \langle w\theta \rangle \right] , \tag{S6.1}$$

$$\langle v^2 \rangle = \frac{E_{\mathrm{k}}}{3} + \frac{l_1}{E_{\mathrm{k}}^{1/2}} \left[2\langle uw \rangle \frac{\mathrm{d}U}{\mathrm{d}z} - 2\frac{g}{\vartheta_{00}} \langle w\theta \rangle \right] , \tag{S6.2}$$

$$\langle w^2 \rangle = \frac{E_{\mathrm{k}}}{3} + \frac{l_1}{E_{\mathrm{k}}^{1/2}} \left[2\langle uw \rangle \frac{\mathrm{d}U}{\mathrm{d}z} + 4\frac{g}{\vartheta_{00}} \langle w\theta \rangle \right] , \tag{S6.3}$$

$$\langle uw \rangle = \frac{3l_1}{E_{\mathrm{k}}^{1/2}} \left[-\left(\langle w^2 \rangle - C_1 E_{\mathrm{k}} \right) \frac{\mathrm{d}U}{\mathrm{d}z} + \frac{g}{\vartheta_{00}} \langle u\theta \rangle \right] , \tag{S6.4}$$

$$\langle u\theta \rangle = \frac{3l_2}{E_{\mathrm{k}}^{1/2}} \left[-\langle uw \rangle \frac{\mathrm{d}\Theta}{\mathrm{d}z} - \langle w\theta \rangle \frac{\mathrm{d}U}{\mathrm{d}z} \right] , \tag{S6.5}$$

$$\langle w\theta \rangle = \frac{3l_2}{E_{\mathrm{k}}^{1/2}} \left[-\langle w^2 \rangle \frac{\mathrm{d}\Theta}{\mathrm{d}z} + \frac{g}{\vartheta_{00}} \langle \theta^2 \rangle \right] , \tag{S6.6}$$

$$\langle \theta^2 \rangle = -\frac{\Lambda_2}{E_{\mathrm{k}}^{1/2}} \langle w\theta \rangle \frac{\mathrm{d}\Theta}{\mathrm{d}z} , \tag{S6.7}$$

$$\varepsilon = \frac{E_{\mathrm{k}}^{3/2}}{\Lambda_1} . \tag{S6.8}$$

Note that Mellor and Yamada (1982) use a symbol (q^2 in that paper) which is twice E_{k}. Accordingly, the lengths here are different from those by Mellor and Yamada (1982), and here $C_1 = 0.16$.

References

C.W. Gardiner, *Handbook of Stochastic Methods for Physics, Chemistry and the Natural Sciences*, 2nd edn. (Springer, Berlin, 1990)

I.S. Gradshteyn, I.M. Ryzhik, *Table of Integrals, Series, and Products*, 1980th edn. (Academic Press, New York, 1980)

G.L. Mellor, T. Yamada, Development of a turbulence closure model for geophysical fluid problems. Rev. Geophys. Space Phys. **20**, 851–875 (1982)

A.S. Monin, A.M. Yaglom, *Statistical Fluid Mechanics*, vol. II (MIT Press, Cambridge, 1975), p. 874

H.R. Olesen, S.E. Larsen, J. Hojstrup, Modelling velocity spectra in the lower part of the planetary boundary layer. Bound.-Layer Meteorol. **29**, 285–312 (1984)

M.R. Raupach, R.H. Shaw, Averaging procedures for flow within vegetation canopies. Bound.-Layer Meteorol. **22**, 79–90 (1982)

Index

F. Tampieri, *Turbulence and Dispersion in the Planetary Boundary Layer*,
Physics of Earth and Space Environments, DOI 10.1007/978-3-319-43604-3

MIX
Papier aus verantwortungsvollen Quellen
Paper from responsible sources
FSC® C105338

If you have any concerns about our products,
you can contact us on
ProductSafety@springernature.com

In case Publisher is established outside the EU,
the EU authorized representative is:
Springer Nature Customer Service Center GmbH
Europaplatz 3, 69115 Heidelberg, Germany

Printed by Libri Plureos GmbH
in Hamburg, Germany